P9-DDG-421

Three Laws of Nature

Three Laws of Nature

A Little Book on Thermodynamics

R. Stephen Berry

Yale UNIVERSITY PRESS
New Haven & London

Published with assistance from the foundation established in memory of
Philip Hamilton McMillan of the Class of 1894, Yale College.

Yale University Press books may be purchased in quantity for educational, business, or promotional use. For information, please e-mail sales.press@yale.edu (U.S. office) or sales@yaleup.co.uk (U.K. office).

Set in Janson Roman type by Integrated Publishing Solutions.
Printed in the United States of America.

ISBN 978-0-300-23878-5 (hardcover : alk. paper)
Library of Congress Control Number: 2018954948
A catalogue record for this book is available from the British Library.

This paper meets the requirements of ANSI/NISO Z39.48-1992
(Permanence of Paper).

10 9 8 7 6 5 4 3 2 1

Contents

Preface

This book has a very specific purpose: to use the science of thermodynamics as a paradigm to show what science is, what science does and how we use it, how science comes to exist, and how science can evolve as we try to reach and address more and more challenging questions about the natural world. It is intended primarily for people with little or no background in science, apart, perhaps, from some exposure in school or college to "science for everyone" courses.

The approach, quite frankly, came from three stimuli. One was a course I gave for non-scientist undergraduates at the University of Chicago that evolved over years. The second was an adult education course that grew out of that undergraduate course—but was at a somewhat lower level, primarily in terms of the mathematics. The third, first in time, which had a very crucial influence on that evolution, was a pair of essays published in 1959 by the British scientist and novelist C. P. Snow, titled "The Two Cultures," based on lectures he had given at Cambridge University. Snow said, in essence, that those who belong to the culture of scientists know vastly more about the culture of humanities than the reverse, and that this is a serious problem for our society. He argued, metaphorically, that if we were to achieve a proper balance, the non-scientists would know as much about the second law of thermodynamics as the scientists know about Shakespeare. In a later essay, actually published in 1964 with the first two, he retreated and replaced the second law of thermodynamics with contemporary biology.

I might have disagreed with his change even then, but as biology has advanced it is very clear that, in our time, he was right in the first place. Why? To understand biology today, one must learn many, many facts, but one needs very few facts to understand thermodynamics. It is a subject rooted in a few concepts, far more than on systematic inferences from vast amounts of information, as contemporary biology is now.

This book is written for people who are interested in what science is and does, and how it comes to be, but who have little or no prior substantive knowledge of any science. In writing for this audience, there was a clear choice to be made regarding the order of the chapters. Would the ideas be clearer if the history came first, or if the first thing one read was the substance of the science as it is today? I have chosen the latter. Thus I first show, in Chapters 1 and 2, what the science of thermodynamics is as we now use it almost constantly, and then, in Chapter 3, address the question of how we arrived where we are.

The first two chapters develop an overview of what the traditional science of thermodynamics is, of the concepts that embody it, and how it uses these concepts. These chapters are built around the three laws of thermodynamics, namely the conservation of energy, the inevitability and direction of change with time, and the existence of a limiting absolute zero of temperature.

The third chapter presents the history of the science. There we see how the concepts, the variables, and the tools that we call the laws of thermodynamics evolved. This is important because some of the concepts of thermodynamics are so ubiquitous and pervasive in our thinking and language that they seem obvious and trivially simple to us, where, in reality, they evolved only with great difficulty, and remain in many ways amazingly subtle and not so simple at all. That history has been, and even still is, a tortuous but fascinating path, especially when we understand where and what thermodynamics is today. And it offers a rich insight into how a science evolves, often having to resolve competing, conflicting concepts.

The fourth chapter describes the ways we apply thermodynamics, especially how we use information from this science to improve the performances of many things we do in everyday life. The fifth chapter describes the way the science of thermodynamics has continued to evolve since its pioneers established its foundations. This is, in a sense, a contemporary counterpart to Chapter 3. Despite the elegant and in many ways the most natural presentation of the subject, in a formal, postulational and essentially closed format, the subject has continued to undergo evolutionary changes, expanding and strengthening as new discoveries require that we extend our concepts. A crucial advance discussed in this chapter is the connection between the macroscopic approach of traditional thermodynamics and the microscopic description of the world, based on its elementary atomic building blocks. This link comes via the application of statistics, specifically through what is called "statistical mechanics."

The sixth chapter examines the open challenges that we face if we wish to build upon the thermodynamics we now have to provide us with tools to go beyond, to other, still more difficult and usually more complex problems of how nature functions. It addresses questions of whether and when there are limits to what a particular science can tell us, and whether inferences we make using the tools and concepts of a science can sometimes have limits of validity. This chapter reveals how this science continues to offer new challenges and opportunities, as we recognize new questions to ask.

The seventh and final chapter addresses the question of how thermodynamics, one particular example of a kind of science, can show us what a science is and does, and hence why we do science at all. This chapter is a kind of overview, in which we look at how thermodynamics can serve as a paradigm for all kinds of science, in that it reveals what scientific knowledge is, how it differs from knowledge in other areas of human experience, how scientific knowledge, as illustrated by thermodynamics, has become deeply integrated into how we live, and how we can use scientific knowledge to help guide human existence.

A comment directly relevant to the intent and content of this book: "Thermodynamics is the science most likely to be true." This is a paraphrase of a comment made by Albert Einstein; he was putting thermodynamics in the larger context of all of science. Einstein certainly recognized that every science is always subject to change, to reinterpretation, even to deep revision; he himself was responsible for such deep changes in what had been accepted as scientific laws. His comment reflects his belief that thermodynamics contains within itself aspects—concepts and relations—that are unlikely to require deep revision, unlike the way quantum theory revised mechanics and relativity revised our concepts of space and gravitation. Was he right?

I would like to acknowledge the very helpful comments from Alexandra Oleson and Margaret Azarian, both experienced and skilled at editing.

Three Laws of Nature

ONE

What Is Thermodynamics?

The First Law

Thermodynamics occupies a special niche among the sciences. It is about everything, but in another sense it is not about anything truly real—or at least, not anything tangible. As far as we know, it applies to everything we observe in the universe, from the smallest submicroscopic particles to entire clusters of galaxies. Yet in its classic form, thermodynamics is about idealized systems: systems that are in a state of equilibrium, despite the obvious fact that nothing in the universe is ever truly in equilibrium, in the strictest sense. Everything in our observable universe is constantly changing, of course—but much of this change happens so slowly that we are unaware of it. Consequently, it is often very useful—and even valid—to treat a system that is not changing in any readily observable way as if it were in equilibrium. Here, traditional thermodynamics shows its full power and allows us to use its concepts to describe such a system.

At this point, we need to be specific about what we mean by "equilibrium," and a state of equilibrium. In thermodynamics, we usually treat systems of macroscopic size—ones that are large enough to see, at least with a magnifying glass, or vastly larger—but not systems on an atomic scale, or even thousands of atoms. (Strictly, we can use thermodynamics to treat macroscopic assemblages of very small things—collections of many, many atoms or molecules, for example—but not those small things individually.) We deal with properties of these macroscopic systems that we can observe and measure, such as temperature, pressure, and volume. We specifically do not include the motions and collisions of the individual molecules or atoms that constitute the systems. After all, thermodynamics developed before we understood the atomic and molecular structure of matter, so we can use its concepts without any explicit reference to the behavior of those constituent particles.

This allows us to give meaning to equilibrium, and states of equilibrium. If the properties we use to describe the state of a macroscopic system, such as its temperature, its composition, its volume, and the pressure it is subject to, remain unchanged for as long as we observe that system, we say it is in equilibrium. It may be in equilibrium only for a matter of minutes, or hours, but if, for example, it is a system in the outdoor environment subject to changes in temperature from day to night, then we might say the system is in equilibrium for minutes or perhaps a few hours, but its state changes as night sets in and the temperature of the surroundings—and hence of the system, too—drops. So in this example, equilibrium is a state that persists only so long as the key properties remain unchanging, perhaps for minutes or hours. But during those time periods, we say the system is in equilibrium. Even when the temperature of the system is changing but only very, very slowly, so slowly that we cannot measure the changes, we can treat the system as if it is in equilibrium. Alternatively, imagine that our observation takes months, so that we see only an average temperature, never the daytime or nighttime temperature. This long-term observation shows us a different kind of equilibrium from what our minute-long or hour-long observation reveals. The key to the concept of equilibrium lies in whether we are unable to see any changes in the properties we use to characterize the macroscopic state of the system we are observing, which, in turn, depends on the length of time our observation requires. Thermodynamics typically involves measurements that we can do in minutes, which determines what we mean by "equilibrium" in practice.

Of course this science has evolved to say some extremely important things about phenomena that are not in equilibrium; as we shall see later, much of the evolution of the subject in recent times has been to extend the traditional subject in order to describe aspects of systems not in equilibrium. But the thermodynamics of systems not in equilibrium falls outside the classic textbook treatment and, in a sense, goes beyond the main target of our treatment here. In the last part of this work, we shall deal with situations out of equilibrium as a natural extension of the subject, in order to illustrate how new questions can lead thinking and inferences into new areas, thereby extending the science.

There is an interesting paradox associated with thermodynamics as a science of systems in equilibrium, because the essential information we extract using traditional thermodynamics centers on what and how changes can occur when something goes from one state—one we can call an equilibrium state—to another such state of equilibrium. It is, in a way, a science of the rules and properties of change. Thermodynamics provides strict rules about what changes can occur and what changes are not allowed by nature.

The first idea that comes to mind for most people when they hear the word "thermodynamics" is that it's about energy. And that is certainly correct; energy is an absolutely central concept in thermodynamics. The first rule one encounters on meeting this subject is a law of nature, known as the first law of thermodynamics, which states that energy can be neither created nor destroyed, but only transformed from one form to another. We shall have much to say about this law; examining the process of its discovery tells us much about the nature of science, about scientific understanding, and about energy itself.

Thermodynamics uses a set of properties to describe the states of systems and the changes between these states. Several of these are commonplace concepts that we use in everyday experience: temperature, pressure, and volume are the most obvious. Mass, too, may appear. So may composition; pure substances and mixtures don't have exactly the same properties because we can typically vary the ratios of component substances in mixtures. The relative amounts of different substances in a mixture becomes a property characteristic of that mixture. Other commonplace properties, such as voltage and current, can sometimes appear when we describe the thermodynamics of an electrical system, and there are similar concepts for magnetic systems.

But there are also some very special concepts—we can now call them properties—that one may not encounter as readily outside the context of thermodynamics or its applications. The most obvious of these is *entropy*, which we shall have much to say about in Chapter 2. Entropy is the central concept associated with the second law of thermodynamics, and is the key to the direction things must move as time passes. (It is also now a key concept in other areas, such as information theory.) There are still other very powerful properties we shall examine that tell us about the limits of performance of systems, whether they be engines, industrial processes, or living systems. These are the quantities called "thermodynamic potentials." Then there is one more law, the third law of thermodynamics, which relates temperature and entropy, specifically at very, very low temperatures.

In this introductory chapter and the next, we examine the laws of thermodynamics and how these are commonly used for a variety of practical purposes as well as for providing fundamental descriptions of how the universe and things in it must behave. In carrying this out, we shall frequently refer to the many observable (and measurable) properties as *variables*, because thermodynamics addresses the way that properties vary, and how they often vary together in specific ways. It is those ways that form the essential information content of the laws of thermodynamics.

The Most Frequently Encountered Variables and the First Law

Thermodynamics approaches the description of nature using properties of matter and how they change, in terms we encounter in everyday experience. In some ways, this description is very different, for example, from the one we associate with Newton's laws of motion or the quantum theory. The laws of motion are the natural concepts for understanding very simple systems such as billiard balls, or the behavior of matter at the size scale of atoms and molecules. We refer to that approach as a *microscopic* description, while thermodynamics we say is at a *macroscopic* level. A thermodynamic description uses properties, mostly familiar ones, that characterize the behavior of systems composed of many, many individual elements, be they atoms, molecules, or stars making up a galaxy. As we said, among them are temperature, pressure, volume, and composition, the substances that constitute the system. In contrast, properties or variables appropriate for a microscopic description include velocity, momentum, angular momentum, and position. Those are appropriate for describing collisions of individual atoms and molecules, for example. We simply can't imagine wanting to keep track of the motion of all the individual molecules of oxygen and nitrogen in the air in a room. The key to a thermodynamic description expresses the changes in the macroscopic properties or variables. The variables we use in thermodynamics (and in some other approaches as well) describe phenomena on the scale of our daily experience, not normally at the atomic level.

Consider how we use those macroscopic variables to describe individual macroscopic objects, such as a baseball; we can determine its volume and temperature, for example. However, when we treat a baseball as an individual thing, as a sphere to be driven by colliding with a bat, paying no attention to its microscopic, atomic-level structure, we can use the same variables we would use at the micro level, mass, position, velocity, angular momentum. The important distinction here is not the size of the individual objects but how simplistic our description of the object is; are we concerned with how individual components that make up the baseball behave and interact, or with the way the system composed of those many individual components behaves as a whole?

There is, however, one property, energy, that is appropriate and important at both levels, so it holds a very special place in how we look at nature. Because we study its changes, we often refer to it as a variable. When we think of energy, we inevitably think of temperature and heat. Temperature is almost universally abbreviated as T, and we shall very often use that symbol. At this point, however, we shall not try to be any more precise about what temperature *is* than to say it is the *intensity of heat*, as measured by a thermometer, that people use different scales to measure it, and that it is how we quantify the intensity of heat.

Obviously we are all familiar with the Fahrenheit scale, in which water freezes at 32° F and boils at 212° F—when the pressure on that water has the precise value we call 1 atmosphere. (This scale was introduced in the early eighteenth century by Daniel Gabriel Fahrenheit, in the Netherlands.) Most readers will also be familiar with the Celsius, or centigrade, scale, in which water freezes at 0° C and boils at 100° C, again at a pressure of precisely 1 atmosphere. (It was Anders Celsius, a Swedish astronomer, who introduced this scale in 1742.) The first letter of the name of the temperature scale, capitalized, denotes the units and scale being used to measure the degrees of temperature—F for Fahrenheit, C for Celsius, and at least one other we'll encounter (which will give us a more precise and less arbitrary way of defining temperature). Temperature is a property unrelated to the amount of material or of space, so long as we consider things that consist of many atoms or molecules; so is density, the mass per unit of volume. We refer to such properties as *intensive*, in contrast to such properties as total mass or volume, which do depend directly on the amount of stuff. Those are called *extensive* properties.

At this point, pressure is a bit easier than temperature to define precisely: it is the force per unit of area, such as the force exerted by 1 pound of mass resting on 1 square inch of area under the attraction of the earth's gravity at an altitude of sea level. Likewise, we can use the force per unit area exerted by the atmosphere on the earth's surface at sea level as another measure of pressure. Volume is still easier: it is simply the amount of space occupied by whatever concerns us, and it can be measured, for example, in either cubic inches or cubic centimeters.

Now we address some slightly more subtle properties that serve as variables in thermodynamics. We said that temperature is a measure of the *intensity* of heat, but how do we measure the *quantity* of heat? The classic way is in terms of how heat changes the temperature of water: 1 calorie is the quantity of heat required to raise the temperature of 1 cubic centimeter (cc) of water by precisely 1° C. Anticipating, we shall soon recognize that the quantity of heat is a quantity of one form of energy. It is a form that can be contained as a property of an unchanging system, or as a form of energy that passes from one body to another, the amount of energy that is exchanged between a source and a receiver or sink. The quantity of heat is conventionally designated as Q.

Next, we come to the concept that, as we shall see, was one of those that motivated the creation of thermodynamics—namely, *work*. Work, in traditional, classical mechanics, is the process of moving something under the influence of a force. It may be accelerating a car, or lifting a shovel full of dirt, or pushing an electric current through some kind of resisting medium such as the filament of an incandescent light bulb. Whatever that force is, so long as there is anything opposing that force, such as gravity when lifting a shovel or inertia when accelerating a car, energy must be exchanged between the source of the work and the source of the resistance. Thus work is that energy used to make something specific happen. It may include work against friction in addition to work against inertia, as it is with the work of moving a piston in an engine. Work is conventionally abbreviated as W. The explicit forms used to evaluate work are always expressions of particular processes, whether lifting a weight against gravity, pounding a nail into a wooden board, pushing electrons through a filament, or whatever. Work, like heat, is a form of energy that is done by some source on some "sink," whether it be a flying object accelerated by the work done by its engine or a balloon being blown up by a child.

Work is energy exchanged in a very ordered way. Consider a piston in an engine. When work is done to accelerate the piston, all of the atoms of the piston move in concert; the piston may or may not change temperature, but it certainly retains its identity. If, on the other hand, we heat the piston without moving it, the individual atoms that constitute it move more vigorously as the temperature of the piston increases. (We shall return to how this came to be understood in Chapter 3, when we examine the history of the concepts of thermodynamics.) Here, we simply recognize that work is energy transferred in a very orderly way, while heat is energy transferred in a very random, disorderly way. Of course when a real system or machine does work, there are invariably ways that some of the energy put into the working device goes into what we call waste—notably friction and heat losses. Yet in order to move a piston, for example, we must accept that we need to overcome not only its inertia, but also the friction that is inherent between the piston and its cylinder. It is often useful for conceptual purposes to imagine ideal systems in which those "wasteful" and unwanted aspects are absent; such idealization is often a useful way to simplify a physical process and make it easier to understand and analyze. We shall see such idealization introduced when we examine the contributions of the French engineer Sadi Carnot in Chapter 3.

One very important thing to recognize from the outset is that heat and work can both be expressed in the *same units.* They are both forms of energy, expressible in calories, the unit we introduced to measure heat, or in ergs, the basic small unit for work in the metric system, or any of several other units, such as joules. (One erg is sometimes said to be "one fly push-up." One joule is 10,000,000 ergs.) Both heat and work, however, must be in units that are essentially a mass multiplying the square of a velocity v, or a mass m, multiplied by the square of a length unit l, and divided by the square of a time unit t. We can express this tersely thus, using square brackets to indicate "the dimensions of": $[Q] = [W] = [ml^2/t^2]$. In this context, "dimension" refers to a physical property. The dimensions of this kind that we encounter most frequently are mass, length, and time, but there are some others, such as electric charge. Alternatively, we can express these dimensions in terms of a velocity v, because velocity is distance per unit of time, or distance/time, l/t. Hence [Q] = [W] = [mv^2]. When we use those dimensions, we give quantitative units for them, such as centimeters or inches for length, grams or kilograms or pounds for mass. Velocity involves two basic dimensions, length and time, so it is measured in units such as miles per hour or centimeters per second.

Heat and work are the two forms of energy that are always associated with some process, and hence are called *process variables*. Temperature, pressure, mass, and volume, on the other hand, describe states of systems (in equilibrium, in principle), and hence are called *state variables*. We shall be using and relating these concepts shortly.

For two quantities to be comparable, they must be expressed in the same dimensions, and, of course, to make that comparison, the quantities must be expressed in the same units, the same measurement scale. We can compare lengths if they are expressed in inches or meters, but if some are in inches and others are in meters or light-years, we cannot compare them until we convert them all into a common set of units. Each dimension has its own set of units. Length has many choices, such as inches, centimeters, or kilometers. Velocity, with dimensions of distance divided by time, or [l/t], can be, for example, miles per hour or meters per second. There are several different units that we use for energy: calories, ergs, joules, for example, each of which can be expressed as some mass unit, multiplying the square of some length unit, divided by the square of some unit measuring time. We will address these options later, as we deal with specific examples.

Now we notice something very important. You can recall from high school science that the *kinetic energy* of a moving object with mass m, moving at a velocity v, is precisely ½ mv^2. This is a hint toward the very fundamental concept that heat, work, and kinetic energy share a common basic characteristic: they are all *forms of energy*. The grand, unifying concept that encompasses these and many other things we can experience or recognize is *energy*. In the third chapter, we shall see how this concept evolved through many years of experiment and thought and even controversy. (In fact, it is even continuing to evolve and expand now, as you read this.) Here, we shall assume that energy is such a widely used concept that the reader has an intuitive sense of what it is. How precise or sophisticated that understanding may be is not important at this point.

The most important of energy's characteristics here is that it can appear in many, many forms. Energy is stored in chemical bonds in coal or natural gas in a form that can be transformed in part into heat. This is achieved simply by subjecting the coal or gas to a chemical reaction, namely combustion—simply burning—which replaces the original chemical bonds between carbon atoms or carbon-hydrogen bonds with new, stronger bonds to oxygen that weren't present in the original fuel. In this way, we transform some chemical energy into heat. Likewise, we can transform some heat energy into electrical energy by using heat to make steam, and then pushing the steam, at fairly high pressure, against the blades of a turbine that rotates a coil of wire through the field of a magnet, a process that yields electric current. This is a rather subtle, complex process, because energy that is heat at our starting point becomes the energy that transforms liquid water to steam and also heats it further; then that energy becomes mechanical energy that rotates the turbine blades, which in turn becomes electrical energy in the form of a current moving at some chosen voltage. This everyday phenomenon is actually a remarkable demonstration of some of the many forms that energy may take.

Here we can introduce the first law of thermodynamics, a phenomenon of nature that is so familiar that we often fail to recognize how astonishing it is. The first law tells us that energy can never be created or destroyed, but only transformed from one form to others. We say that energy is *conserved.*

First, however, we need to say a little more about those variables of energy, heat, and work. If a system, any system, is in an unchanging equilibrium state, *and is isolated from interacting with anything else*, then certain of its properties are also unchanging. One of those properties is the energy of that system. The energy of an isolated system is a property, but, because of that isolation, it cannot be changed; it is a constant, one that at least partially characterizes the properties of that system, so long as the system cannot interact with anything outside itself. We can think of our entire universe as such a system; so far as we can tell, there is nothing else with which it can interact, and the principle of conservation of energy says that therefore the total energy of the universe cannot change. We frequently idealize small systems, such as objects we can use for experiments, as being isolated from their environment; thus, for practical purposes, we can, for example, wrap enough insulation around an object that for a long time, it effectively has no contact or interaction with its surroundings. That way, it cannot exchange energy with those surroundings, whether losing or gaining it. In such a case, that isolated object must have an unchanging amount of energy. For isolated systems, therefore, we say that *energy is a property of the state of the system*, or *energy is a state variable.* Every isolated system has a precise energy of its own, an energy that can only change by spoiling that isolation and allowing the system to interact—more precisely, to take in or give off energy—with whatever other system it is interacting with.

We can envision other kinds of states of equilibrium, states in which nothing changes with time. One of the most useful is the state of a system that is in contact with a thermostat, a device, perhaps idealized, that maintains that system at a constant temperature T. The temperature is a property of the state of this system, and hence is also a *state variable.* The mass m, the volume V, and the pressure p are also state variables. Some state variables, such as mass and volume, depend directly on the size of the system; double the size of the system and you double its mass and volume. These are called *extensive variables.* Others, typically temperature, density, and pressure, do not depend at all on the size of the system; these are called *intensive variables.* Color is another intensive characteristic, but is rarely relevant to thermodynamics—although it may be important for energy transfer, as with absorption of light.

Now we turn to work and heat. Work is something that we do, or a system does, when it changes its state and, typically, its environment as well. Likewise heat is a form of energy that a system exchanges with its surroundings, either absorbing it or depositing it. Neither work nor heat is a state variable; they are both what we can call process variables. They are the variables describing the way the system undergoes change, from one state to another. When a system changes its state, it does so by exchanging work, heat, or both with its surroundings. And this distinction, between state variables and process variables, brings us to an effective way to express the first law of thermodynamics. Work and heat are the only two kinds of process variables that we know and use. Of course there are many kinds of work—mechanical, electrical, and magnetic, for example. Although heat may pass from warmer to cooler bodies in various forms, we know only one kind of heat. Work is energy being transferred from a source into a very specific form, such as the motion of a piston or the initiation and maintenance of an electric current. Heat is perhaps best understood by thinking of it at a microscopic level; at the macroscopic level, the amount of heat in a system determines the temperature of that system. However at the microscopic level, the heat contained in a system determines the extent to which the component atoms and molecules move, in their random way. The more heat, the more vigorous the motion of those tiny particles.

As a matter of convention, we will say that if heat is added to the system, then the quantity of heat Q is positive. Likewise, if the system does work, we will say that the amount of work W is positive. Hence positive Q increases the energy of the system, and positive W takes energy from the system and thereby decreases its energy. As an aside, we note that the convention for Q is the one generally accepted. For the sign of W, however, both conventions are used by different authors; some take positive W as work done *on* the system and hence increasing its energy. Either convention is consistent and workable; we just have to decide which we will use and stay with that.

Now we need to introduce some notation, in order to express the first law and its consequences. We shall use the standard symbol E to represent energy, specifically the amount of energy. Another standard representation we shall use is the set of symbols generally taken to signify changes. To specify changes, it is standard practice to use letters from the Greek alphabet. We denote very small changes in any quantity, call it X, with the letter *d* or a Greek δ (lowercase delta), either dX or δX. These typically can be so small that an ordinary observation would, at best, barely detect them. We designate a very small change in pressure p as dp or δp. Significant, observable changes in X we indicate with a Greek Δ (uppercase delta), as ΔX. We designate a large change in the volume V of a system as ΔV. With this bit of notation, we move toward stating the first law in a form that makes it extremely useful.

The conservation of energy and the identification of energy as a state variable for an isolated system tells us that if a system goes from some state, say #1, to another state, say #2, then the change in the system's energy is simply the difference between the energies of states 1 and 2. We can write this as an equation:

$$\Delta E = E(2) - E(1).$$

The important point we must now recognize is that this change in energy doesn't depend in any way on how we bring the system from state 1 to state 2. We might be able to make the change by heating or cooling alone, or by doing or extracting work with no exchange of heat with an outside source, or by some combination of the two. No matter how we take the system from state 1 to state 2, the energy change is the same. This means that however we exchange heat or perform or extract work, *their difference must be the same fixed amount, the difference between* E(1) and E(2). We can write this as an equation, either for measurable changes of energy, as

$$\Delta E = Q - W$$

or in terms of even the smallest changes, as

$$\delta E = \delta Q - \delta W.$$

These equations say that however we put in or take out heat, and however the system does work or has work done on it, the difference between the heat taken in and the work done is precisely equal to the change in energy of the system from its initial to its final state. It is these equations that are our terse statement of the first law of thermodynamics. These two equations tell us that however much work we do or extract to take our system, say our engine, from its initial state 1 to its final state 2, there must be an exchange of heat corresponding precisely to that amount of work so that the change of energy is just that difference between the two states. Said another way, these equations tell us that there is no way to create or destroy energy; we can only change its form and where it is.

One very simple example illustrates this principle. The energy of a 20-pound weight is higher if it is on top of a 50-foot pole than if it is on the ground at the base of the pole. But the difference between the energies of that weight does not depend in any way on how we brought it to the top of the pole. We could have hoisted it up directly from the ground, or dropped it from a helicopter hovering at an altitude of 100 feet; the difference in the energy of the ground-level state and the state 50 feet higher in no way depends on how the states were produced.

This also tells us something more about how real machines operate. Think of an engine such as the gasoline engine that drives a typical automobile. Such an engine operates in *cycles*. Strictly, each cylinder goes through a cycle of taking in fuel and air, igniting and burning the fuel, using the heat to expand the gas-air mixture, which pushes the moveable piston of the cylinder. The moving piston connects via a connecting rod to a crankshaft and the expansion in the cylinder, pushing the piston, in turn pushes the crankshaft to make it rotate. That crankshaft's rotation eventually drives the wheels of the car. The important part of that complex transformation for us at this point is simply that each cylinder of the engine goes regularly through a cycle, returning to its initial state, ready to take in more fuel and air. In the process of going through the cycle, the burning of the gasoline in air releases energy stored in chemical bonds and turns it into heat. The engine turns that heat into work by raising the pressure of the burned gasoline-air mix enough to push the piston through the cylinder. At the end of each cycle of each cylinder, the piston returns to its original position, having converted chemical energy to heat, and then, transformed that heat into work.

In an ideal engine, with no friction or loss of heat to the surroundings, the engine would return to its initial state and as much as possible of the chemical energy that became heat would have become work, turning the wheels of the car. ("As much as possible" is a very important qualifier here, which will be a central topic of a later discussion.) Such an engine, first envisioned by the French engineer Sadi Carnot, has the remarkable property that at any stage, it can move equally easily in either direction, forward or backward. We call such an ideal engine or process a *reversible engine* or *reversible process.* One extremely important characteristic of reversible processes is that they can only operate infinitely slowly. At any instant, the state of a reversible engine is indistinguishable from—or equivalent to—a state of equilibrium, because it can go with equal ease along any direction. Such reversible processes are ideal *and unreachable* limits corresponding to the (unreachable) limits of the best possible performance that any imaginable engine could do. However, we can describe their properties and behavior precisely because they are the limits we can conceive by extrapolating the behavior of real systems. They would operate infinitely slowly, so obviously are only useful as conceptual limits, and in no way as anything practical.

Of course real engines are never those ideal, perfect machines. They always have some friction, and there is always some leakage of heat through the metal walls of the engine to the outside. That the engine feels hot when it has been running testifies to the loss of heat to the environment; that we can hear the engine running is an indication that there is friction somewhere that is turning some, maybe very little, of that energy that was heat into sound waves, generated by friction. But we try to minimize those losses by using lubricating oil, for example, to keep the friction as low as we can. We make real engines that are as much like those ideal engines as we can, but of course require the real engines to operate at real rates, not infinitely slowly, to deliver the work we want from them.

One very important, inherent characteristic of what we call *energy* is the remarkably broad scope of what it encompasses, as we already have begun to see. Every activity, every process, even every *thing* there is has associated with it an energy or energy change. The concept of energy includes, in its own way, some aspect of virtually everything in human experience. In the next chapter, we shall examine how this concept evolved and grew, as human experience and understanding of nature grew. At this stage, we need only recognize that energy is a property we can identify, often measure, and detect in an amazingly broad variety of forms. Light, sound, electricity and magnetism, gravitation, any motion whether constant or changing, the "glue" that holds atomic nuclei together, even mass itself, are all examples of the manifestation of energy. And now, there are even forms of energy that we do not yet understand, what we call "dark matter" and "dark energy." That the human mind has recognized the very existence of such a universal manifestation of the natural world is itself a source of amazement and wonder.

A Closer Look at Some Basic Concepts for the Next Big Step

First, we consider *temperature*. Strictly, the concept evolved as a descriptive property only of systems in equilibrium. In practice today, it is also applicable in many circumstances, but not all, to systems undergoing transitions from one state to another. All that is required to do this is that the process is slow enough to put in some kind of temperature-measuring device that will give us a number for that place where we inserted our thermometer. Our intuitive sense of temperature is a measure of an intensity, specifically of how much energy a system contains in any part of itself, and intuitively, that form of energy that we recognize as the degree of warmness or coolness. (We choose to use these words, rather than "the degree or intensity of heat" because we will shortly return to "heat" as a related but different concept.) A system in equilibrium has the same temperature everywhere in that system. It doesn't matter whether the system is small or large; temperature is a property that is independent of the size of the system, and hence is an intensive property. We shall say a bit more below about properties and variables that are neither intensive nor extensive.

Heat is the form of energy that most often changes the temperature of a system. Heat, as the term is used in the sciences, is a property associated with the *change of energy* of something, with its change of state, not a property of the state of the system. As we saw, properties or variables that describe changes—gains or losses—are called process variables, in contrast to state variables. The concept of heat went through a fascinating history that we examine in the third chapter. In essence, there were two incompatible ideas of what heat really is. One view saw heat as an actual fluid, which was given the name "caloric"; the other view saw heat as a measure of the degree of intensity of motion of the small particles that constitute matter. This latter concept emerged even before the clear establishment of the atomic structure of matter, although the evolution of the two concepts linked the ideas of heat and atomic structure. (Of course the notion that matter is composed of atoms had its origin in ancient Greece, with Democritus and Lucretius, but the concept remained controversial for centuries.)

Today, of course, we interpret heat and temperature in terms of the motions of the atoms and molecules that are the basic constituents of matter. Temperature is the specific measure of the average motion of each elemental component, and more specifically, of the *random* motion of the atoms and molecules; heat is the form of energy that directly changes that random motion of the atoms and molecules. If, for example, a substance absorbs light of a specific wavelength, then in most typical situations, that light energy is quickly converted from being concentrated in the particular atomic or molecular structure that acts as an antenna to absorb it, into energy distributed evenly, on average, throughout the entire system, in all the kinds of motion and other ways that energy can exist in the system. This is what happens, for example, when sunlight warms your skin; the energy in the visible, ultraviolet, and infrared solar radiation strikes your body, your skin absorbs that radiation and converts it to the motion of the atoms that make up your skin and muscles that then feel warmed. We say that the energy is *equipartitioned*, distributed evenly among all the places it can be, when the system has come to equilibrium. The *average* amount of energy in any part of a system at a specific temperature is the same as in any other connected part of the system. We shall see, however, that it is the average that is constant when the temperature is constant, and that the amount of energy in any small part of a system actually fluctuates in a random fashion, but with a specific probability distribution.

The other property that can express a change of energy, and hence a change of the state of a system, is *work*. If a piston moves to push a crankshaft, then all the atoms of that piston move together to achieve that push. If there is any resistance to the pushing force, then energy must be exchanged between the piston and whatever is resisting the push. That energy must come from the system of which the piston is a part, and must go to that other system resisting the push. (Think of the inertia and friction that must be overcome by whatever energy a source supplies.) In this situation, as whenever work is done, energy is exchanged, but in a very non-random way. Work is, in that sense, completely different from heat, a complementary way for energy to pass out of or into a system. Heat is energy exchanged in the most randomly distributed way possible; work is energy exchanged in the least randomly distributed way possible. In real systems, of course, there are always frictional and other "losses" that prevent the complete conversion of energy from its initial form into the intended work. Nevertheless, we can imagine ideal engines in which those losses disappear, and build our theoretical ideas with those ideal examples as the natural limits of real systems, as we make them better and better. The ideal reversible engine is that kind of ideal system.

We describe the forms in which energy can be stored and transferred through the concept of *degrees of freedom.* Motion of an object in a specific direction is kinetic energy in a single degree of freedom, the degree corresponding to motion in that direction. A tennis ball can move in three independent directions in space, which we identify with three degrees of freedom, specifically three *translational* degrees of freedom. If it were confined to stay on a plane, on the ground of the tennis court, it would have only two degrees of freedom of translation. But the tennis ball can spin, too; it can rotate about three different axes, so, being a three-dimensional object, it also has three *rotational* degrees of freedom. A typical molecule, composed of two or more atoms, can exhibit vibrations of those atoms as well. The atoms can oscillate toward and away from one another, still retaining the basic structure of the molecule they form. Hence there are *vibrational* degrees of freedom. Molecules, like tennis balls, can rotate in space, and thus have rotational degrees of freedom as well. Every free particle, every atom, brings with it its own three translational degrees of freedom. (Here, we neglect the possibility that such a free particle might also be capable of rotation.) When these atoms stick together to make molecules or larger objects, the nature of the degrees of freedom changes but the total number of degrees of freedom, the number of ways or places where energy can go, remains the same—three times the number of particles that make up the system.

A complex object such as a molecule made of n atoms has 3n degrees of freedom total, but only three of those degrees are degrees of translational freedom, describing the motion of its center of mass. This is because those n atoms are bound together into a relatively stable object. Besides its three translations, the molecule has three degrees of rotational freedom, and all the rest, 3n - 6, are vibrational. As an example to illustrate this, think of two hydrogen atoms and an oxygen atom. If they are all free, each atom has three translational degrees of freedom, nine in all; if, on the other hand, they are bound together as a water molecule, two hydrogens attached to the oxygen, this system has only three translational degrees of freedom, corresponding to motion of its center of mass in the x, y, and z directions; however, it has three rotational degrees of freedom, rotation about each of those three directions, and three vibrational degrees of freedom, a symmetric stretching motion of the O-H bonds, an asymmetric stretching motion with one bond extending while the other shrinks, and a bending vibration. Hence the bound system has the same number of degrees of freedom as the three free particles, but of different kinds. We shall use this concept in several ways in the following discussion.

Some state variables such as temperature and pressure are, as we just saw, independent of the size of the system; these are called *intensive* variables. The other variables most frequently encountered, such as mass, volume, energy, and entropy, depend directly on the size of the system, the amount of stuff that composes that system. These are called *extensive variables.* Together, these are the variables we use to describe systems that, for heuristic purposes, we can validly treat as being in states of equilibrium, whose state variables do not change on time scales such that we can observe them. (There is at least one other kind of variable, exemplified by gravitational energy, discussed below, which depends on the product of the masses of two interacting bodies, and hence is *quadratic* in its dependence on the amount of matter.)

Much of the thermodynamics we encounter in everyday life can be encompassed entirely with intensive and extensive variables. However, if we wish to use thermodynamics to describe the behavior of a cloud of galaxies, for example, for which a large fraction of the energy is gravitational, we can't use just that simple pair of characteristics because gravitational energy depends on the product of the masses of pairs of attracting bodies, such as the earth and the sun. This means that energy associated with gravitation depends on the product of two interacting masses, not simply on a single mass, as with an extensive variable. Hence astronomers wanting to use thermodynamics to portray such things as galaxies and solar systems have had to develop a form of the subject called *nonextensive thermodynamics* to address their systems. Here, however, we shall be able to stay almost entirely within the scope of conventional thermodynamics and its two kinds of variables.

Summary

We have seen how the first law of thermodynamics can be expressed in words: energy can never be created or destroyed but only changed in form and location. We have also seen how this law can be expressed in a very simple equation,

$$\Delta E = Q - W$$

which tells us that ΔE, the change in energy of any system, is precisely the heat Q that the system takes in, minus the work W that the system does on something outside itself. It says that the change of energy of the system, from an initial state to some final state, depends only on the difference of the energies of those two states and not on the pathway taken to go from one to the other. It says that if a system traverses a closed loop, that brings it back to its initial state, that the system has undergone no change in its energy. It is a law we believe is valid everywhere in the universe, a law that describes every form that energy can take. When we realize all the forms that we know energy can have—heat, work, electromagnetic waves, gravitation, mass—this becomes a remarkable, even awesome statement.

TWO

Why We Can't Go Back in Time

The Second and Third Laws

The second law of thermodynamics is very different from the first law. The first law is about something that doesn't change, specifically the total amount of energy, or even the amount of energy in some specific part of the universe, closed off from the rest of the world. The second law is about things that can't happen, which can be interpreted as distinguishing things that can happen from those that can't—which is a way of showing what the direction of time must be.

The first explicit formulations of the second law were three verbal statements of the impossibility of certain kinds of processes. In 1850, the German physicist and mathematician Rudolf Clausius put it this way: no engine, operating cyclically, can move heat from a colder to a hotter body without doing work. Refrigerators can't operate spontaneously. Then, the next year, the British scientist William Thomson, Lord Kelvin, stated it differently: no engine, operating cyclically, can extract heat from a reservoir at a constant temperature and convert that heat entirely into work. The third of these statements, by Constantin Carathéodory in 1909, is, in a sense, more general and less specifically related to work than its predecessors; it says that every equilibrium state of a closed system has some arbitrarily close neighboring states that cannot be reached by any spontaneous process (or by the reversible limit of a spontaneous process) involving no transfer of heat. "You can't get there from here." This, one can see, encompasses the two earlier statements and carries with it the implication that one may be able to go from state A to state B by a spontaneous or limiting reversible process, yet be unable to go from B back to A by such a process. This is a very fundamental way to express the unidirectional nature of the evolution of the universe in time. For example, if state A consists of a warm object X connected to a cooler object Y by some means that can conduct heat, then X will cool and Y will warm until X and Y are at the same temperature—and that is our state B. We well recognize that the system consisting of X, Y, and a connector between them will go spontaneously from the state A in which the temperatures of X and Y are different to the state B, in which X and Y are in thermal equilibrium, but that we will never see the system go from thermal equilibrium to two different temperatures.

At this point, we must introduce a new, fundamental concept

that becomes a most important variable now, for our second law of thermodynamics. This variable is very different from energy, and the natural law associated with it, likewise, is very different from the first law. This new, fundamental variable is *entropy.* Of all the fundamental concepts in thermodynamics, entropy is surely the one least familiar to most people. The concept of entropy arose as the concepts of heat and temperature evolved. More specifically, entropy emerged out of an attempt to understand how much heat must be exchanged to change the temperature of an object or a system by a chosen amount. What is the actual heat exchanged, per degree of temperature change in the system? Then the natural next question is, "What is the absolute minimum amount of heat, for example the number of calories, that must be exchanged to change the system's temperature by, say, 1 degree Celsius?" The most efficient engine imaginable would exchange only the minimum amount of heat at each step of its operation, with none wasted to friction or loss through the system's walls to its environment.

Now, in contrast to those earlier statements, the most widely used expression of the second law is in terms of entropy. Entropy, expressed as a measure of the amount of heat actually exchanged per degree of temperature change, means that it is very closely related to the property of any substance called its *heat capacity*, or, if it is expressed as a property of some specified, basic unit of mass of a material, as *specific heat.* Precisely defined, the heat capacity of a substance is the amount of heat it must absorb to increase its temperature T by 1°. Correspondingly, the specific heat of that substance is the amount of heat required to raise the temperature of one unit of mass by 1°. Typically, both are expressed in units of calories per degree C or K, where K refers to the Kelvin temperature scale, named for Lord Kelvin. The units of measure in this scale, which are called kelvins, not degrees, are the same size as those of the Celsius scale, but zero is at the lowest physically meaningful temperature, a presumably unattainable level at which point absolutely no more energy can be extracted from the system; at 0 K, there is zero energy. This point, which is often referred to as absolute zero, is just below –273° C; hence, 0° C is essentially 273 K. (We shall return to the Kelvin scale when we consider the third law of thermodynamics.) At this point, we need to look at heat capacities and specific heats, in order to probe this approach to the concept of entropy.

The heat capacity of anything obviously depends on its size; a quart of water has twice the heat capacity of a pint of water. To attribute the temperature response of a substance without having to refer to the amount of matter, we use the heat capacity per unit of mass of that substance, which is the *specific heat* of the substance. Thus the specific heat of water is 1 calorie per degree C, per gram or cubic centimeter of water. Specific heat is an intensive quantity (which does not depend on the size of the object); heat capacity is its corresponding extensive quantity, proportional to the mass of the object. Hence the heat capacity of 1 gram of water is the same as its specific heat, but the heat capacity of 5 grams of water is 5 calories per degree C. In the next chapter, we shall discuss the role played by specific heats and heat capacities in the evolution of the concepts of thermodynamics. Typical specific heats range from about 1.2 calories per degree C per gram for helium to much lower values, such as approximately 0.5 for ethyl alcohol, only 0.1 for iron, and 0.03 for gold and lead. Metals require significantly smaller amounts of heat to raise their temperatures.

One subtlety we can mention here is the dependence of the specific heat and heat capacity on the constraints of the particular process one chooses to heat or cool. In particular, there is a difference between heating a system at a constant pressure and heating it while fixing its volume. If the pressure is constant, the volume of the system can change as heat is added or withdrawn, which means the process involves some work, so we must make a distinction between the heat capacities and specific heats under these two conditions.

The *entropy change*, when a substance absorbs or releases energy as heat, is symbolized as ΔS, where S is the symbol generally used for entropy, and Δ, as before, is the symbol meaning "change of." As we discussed, we use Δ to indicate a significant or observable change, while the lowercase δ indicates a very tiny or infinitesimal change—of anything. Then when a system undergoes a change of state by absorbing a tiny bit of heat, δQ, its entropy change δS must be equal to *or greater than* $\delta Q/T$. Formally, we say $\delta S \geq \delta Q/T$. The entropy change must be equal to *or greater than* the amount of heat exchanged, per degree of temperature of the system. Thus, for a given exchange of heat, the entropy change is typically greater at low temperatures than at higher ones. An input of a little heat at low temperatures makes a bigger difference in the number of newly accessible states than at higher temperatures, and it is just that number of states the system can reach that determines the entropy.

That is one quantitative way to state the second law of thermodynamics, in terms of an *inequality*, rather than an equation of equality, such as the statement of the first law. But that doesn't yet tell us what entropy *is*, in a useful, precise way. That only tells us that it is a property that is somehow bounded on one side, that its changes can't be less than some lower limit related to the capacity of a system to absorb or release energy as heat. And that it depends inversely on the temperature at which that heat exchange occurs and hence somehow on the number of ways the system can exist. But from this inequality, we can get a first sense of what entropy tells us. If a substance absorbs an amount δQ of heat and all of that heat goes into raising the temperature of the substance according to that substance's heat capacity, then the process would have an entropy change δS precisely equal to $\delta Q/T$. If some of the heat were to escape by leaking to the surroundings, then the entropy change would be greater than $\delta Q/T$. Hence we can already see that entropy somehow reflects or measures something wasteful or unintended or disordered.

A simple example illustrates how temperature influences entropy. For a gas with a constant heat capacity, the entropy increases with the logarithm of the absolute temperature T. Hence the entropy doubles with each tenfold increase in the absolute temperature; it doubles when the temperature goes from 1 K to 10 K. But it only doubles again when the temperature rises from 100 K to 1000 K! The lower the temperature, the more a given rise in temperature (in degrees) affects and thereby increases the system's entropy. Hot systems are inherently more disordered and thereby have higher entropy than cold systems, so a small increase in the temperature of a hot and already disordered system doesn't make it much more disordered.

Thus far, we have used the word "entropy," and begun to describe a little of its properties, but we have not yet said what entropy *is*. To do that, we will begin to bridge between the macroscopic world of our everyday observations and the microscopic world of atoms and molecules. At our macroscopic level, we describe a *state* of a system in terms of the familiar macroscopic properties, typically temperature, pressure, density, and mass or volume. In fact, any equilibrium state of a macroscopic system can be specified uniquely by assigning specific values to only a subset of these variables; that is enough to determine all the other state variables. That is because the state variables of a substance are linked in a way expressed by an equation, the *equation of state*, unique for each substance. These equations can be simple for simplified models, as we shall see, or they can be very complex if they are put to use to get precise information about real systems, such as the steam in a steam turbine.

In the following discussion, we shall designate a state specified by those state variables as a *macrostate*. The number of those variables that are independent—that is, can change independently—is fixed by what is perhaps the simplest fundamental equation in all of science, the phase rule, introduced by a great American scientist, long a professor at Yale, J. Willard Gibbs, between 1875 and 1878. This rule relates three things. The first, the one we want to know, is the number of degrees of freedom, f, which is the number of things one can change and still keep the same kind of system. This is directly and simply determined by the second quantity, the number of substances or components c in the system, and the third, the number of phases p, such as liquid, gas, or solid, of a specific structure that coexist in equilibrium. These are linked by the relation

$$f = c - p + 2.$$

Not even a multiplication! It's just addition and subtraction. We will later go much deeper into this profound equation. For now, we need only say that every phase of any macroscopic amount of a substance has its own relation, its *equation of state*, that fixes what values all the other variables have if we specify the values of just any two of them.

Typically, there is a different equation of state for the solid, the liquid, and the gaseous form of any chosen substance. (Some solid materials may have any of several forms, depending on the conditions of temperature and pressure. Each form has its own equation of state.) The simplest of all these equations is probably the equation of state for the ideal gas, a sort of limiting case to describe the behavior of gases in a very convenient way that is often a very good approximation to reality. This equation relates the pressure p (not to be confused with the number of phases *p*), the volume V, and the temperature T based on the absolute Kelvin scale, where the lowest point is known as absolute zero. The relation involves one natural constant, usually called "the gas constant" and symbolized as R, and one measure of the amount of material, effectively the number of atoms or molecules constituting the gas, counted not one at a time or even by the thousands, but by a much larger number, the number of *moles.*

A *mole* is approximately 6×10^{23} things, and is essentially an expression of a conversion from one mass scale to another. In the atomic mass scale, a hydrogen atom has a mass of 1—that is, 1 atomic mass unit. Collecting 6×10^{23} hydrogen atoms, we would have 1 gram. Hence we can think of that number, 6×10^{23}, as the conversion factor between atomic mass units and grams, just as 2.2 is the number of pounds in a kilogram. The number 6×10^{23} is known as Avogadro's number, and is symbolized as N_A. The number of moles of the gas is represented as n. To get a feeling for the amount of material in one mole, think of water, H_2O, which has a molecular weight of 18 atomic mass units, so one mole of water weighs 18 grams. But 18 grams of water occupies 18 cubic centimeters, or about 1.1 cubic inches. Hence 18 cubic centimeters of water contains 6×10^{23} molecules, one mole of water.

So now we can write the ideal gas equation of state:

$$pV = nRT.$$

Most other equations of state are far more complicated. The equation of state for steam, used regularly by engineers concerned with such things as steam turbines, occupies many pages. But every state of every substance has its own equation of state. From these equations, we can infer the values of any macroscopic variable we wish from knowledge of any two and, if the variable we want is extensive, of the amount of material. We shall return to this subject in later chapters and derive this equation. (If we wanted to count the particles individually, instead of a mole at a time, we could write the equation as $pV = NkT$, where we use N to indicate the number of atoms or molecules and k is known as the Boltzmann constant, so the gas constant $R = N_A k$, Avogadro's number of little Boltzmann constants.)

At this point we can begin to explore what entropy is. We easily recognize that every macroscopic state of any system, the kind of state specified by the variables of the appropriate equation of state, can be realized in a vast number of ways in terms of the locations and velocities of the component atoms or molecules. All those individual conditions in which the positions and velocities of all the component particles are specified are called *microstates.* Microstates of gases and liquids change constantly, as the component atoms or molecules move and collide with one another and with the walls of their container. In fact, when we observe a macrostate, whatever process we use invariably requires enough time that the system passes through many microstates, so we really observe a time average of those microstates when we identify a macrostate. The molecules of oxygen and nitrogen in the air in our room are of course in constant—and constantly changing—motion, hence constantly changing their microstate. However, we think of the air in a room as being still, at a constant temperature and pressure, and thus in a nonchanging macrostate.

In a microstate, any two identical particles can be exchanged, one for the other. We would call the two states, with and without that exchange, two microstates. Nevertheless we would identify both as representative of the same macroscopic state. The *entropy* of a macroscopic state or *macrostate* is the measure of the number of different microstates that we identify as corresponding to that macrostate. Put more crudely, we can say that entropy is a measure of how many ways all the component atoms could be, that we would say are all the same for us. For example, the microstates of the air molecules in a room are all the possible positions and velocities of all those molecules that are consistent with the current temperature and pressure of the room. Obviously the numbers of those microstates are so vast that, instead of using the actual number of microstates accessible to a system, we use the *logarithm* of that number. Logarithms are very convenient for any situation in which we must deal with very large numbers. The logarithm of a number is the exponent, the power to which some reference number must be raised to produce the given number. The reference number is called the *base*. The three most common bases are 10, 2, and the "natural number" *e*, approximately 2.718. Logarithms to the base 10 are written "log," so 2 is log(100) because $10^2 = 100$. Logarithms to the base *e* are usually written as *ln*, meaning "natural logarithm." Logarithms to the base 2 are often written $\log_2(\ )$ so $\log_2(8) = 3$ because $2^3 = 8$. We shall discuss this in more detail in the context of entropy in the next chapter.

Defined precisely, the entropy of any macrostate is the natural logarithm, the logarithm to the base *e*, of the number of microstates that we identify with that macrostate. Of course the microstates are constantly changing on the time scale of atomic motions, yet we observe only the one constant macrostate when we observe a system in thermodynamic equilibrium. Virtually every macrostate we encounter is some kind of time average over many, many microstates through which our observed system passes during our observation. The oxygen and nitrogen molecules in a room move constantly and rapidly—but randomly—while we sit in what we consider still air at a constant temperature and pressure.

Logarithms are pure numbers, not units of any property.

However, strictly for practical use, we express the entropy in physical units; usually in units of energy per degree of temperature, the same units that express heat capacity. These may be either energy per mole per degree, in which case we multiply the natural logarithm by R, or energy per molecule (or atom, if the system is composed of independent atoms), in which case we use the much smaller constant factor, the Boltzmann constant, symbolized k, the gas constant divided by Avogadro's number, R/N. It is in these units that the entropy was introduced implicitly in the statement of the second law of thermodynamics.

Now we can express the second law in other ways. One is this: All systems in nature evolve by going from states of lower entropy to states of higher entropy—that is, from macrostates of fewer microstates to macrostates of more possible microstates—unless some work is done in the process of changing the state. Another way to say this is that systems evolve naturally from states of lower probability to states of higher probability. A gas expands when its volume increases because there are more ways the molecules of that gas can be in macroscopic equilibrium in the larger volume than in the smaller. To go to a state with fewer microstates, we must do work; we must put in energy.

Another statement of the second law, less general than the above, is this: Systems in contact that are initially at different temperatures but are allowed to exchange energy as heat will eventually come to some common temperature, determined only by the initial temperatures and heat capacities of each of the systems. The composite system relaxes to a common equilibrium by going to the state of highest entropy available to it. That is the state in which the systems share a common temperature. All states in which the systems have different temperatures are at lower entropy.

Here we have crossed a bit from the traditionally macroscopic approach to thermodynamics to the statistically based microscopic approach. While thermodynamics developed originally as a macroscopic view of heat, work, and energy, the introduction of an approach based on statistical mechanics, the statistical analysis of the behavior of the mechanics of a complex system, provided a whole new level of insight and a perspective on the relationship between the traditional approach and our new understanding of the atomistic nature of matter. Statistical mechanics is our first bridge between the macro-level and the micro-level for describing natural phenomena. And it is in the second law that the statistical, microscopic approach gives us the greatest new insight. The world simply moves from the less probable state where it is now to the next, more probable state, the new state which has more available microstates, more ways of being—that is, has a higher entropy—than the present state which it is now leaving. If we want some subsystem, some machine for example, to do otherwise, we must provide energy and do work to make that happen.

Historically, the way entropy was first introduced was, in fact, based on the amount of heat exchanged, per degree of temperature. Specifically, the entropy change is expressed as a limit, based on the ideal, reversible process. If there are just two components, one at a higher temperature T_1 and the other at a lower temperature T_2, then the heat lost from the high-temperature component, Q_1, and that gained by the low-temperature component, Q_2, satisfy the relation that $Q_1/T_1 + Q_2/T_2 = 0$ in that ideal limit. Obviously in a natural flow of heat, Q_1 must be negative and Q_2 positive. The ratio of heat exchanged to the temperature at which that heat is gained or lost is the lower bound on the entropy change in this purely thermal process. If the process is real and not an ideal, reversible one, then the total entropy change must be *greater* than the sum of the ratios of heat exchanged per degree and the sum of heats per degree of temperature must be greater than zero. This macroscopic approach does not offer a way to evaluate the actual entropy change; it only sets a lower bound to its value. Another way to look at that relation is to rearrange the equation to the form $Q_1/Q_2 = -T_1/T_2$. That is, the ratio of heat changes is given by the ratio of the initial temperatures; since T_1 is the higher temperature, the heat lost from that component is greater than the heat gained by the colder component, when the two components come to equilibrium at a common temperature between T_1 and T_2.

In the following chapter we shall investigate the nature of entropy in more detail, as we explore how the concept developed, first at the macroscopic level, and then in terms of behavior at the microscopic level. The essence of the latter, which perhaps

Figure 1. Walter Nernst.
(Courtesy of the Smithsonian Institution Libraries)

gives us the deepest and fullest understanding of the concept, is the intuitively plausible idea that if a system can evolve spontaneously from a macrostate with a smaller number of microstates to one with a larger number of microstates, it will do just that. We can think of that statement of spontaneous behavior, obvious as it seems when expressed in those terms, as one statement of the second law of thermodynamics.

We can conclude this chapter with a statement of the third law of thermodynamics, presented by the German chemist Walter Nernst in 1912 (Fig. 1). This law states that there is an absolute lower bound to temperature, which we call "absolute zero," and that this bound can never actually be reached in a finite number of steps. In practice, experimental advances have enabled us to reach temperatures in the range of levels as low as nanokelvin, or 10^{-9} K. Substances behave very, very differently under those conditions than at the temperatures with which we are familiar. Some, for example, lose all of their electrical resistance, becoming what are called "superconductors." Even more dramatic changes can occur for substances such as helium atoms, which have nuclei that are composed of two protons and two neutrons. At such low temperatures, these atoms can all go into a common quantum state, forming what is called a "Bose-Einstein condensate." Nevertheless, the third law says that absolute zero of temperature remains infinitely far away, at least for macroscopic systems, those we encounter in everyday life.

The concept of an absolute zero of temperature is closely linked to the concept of energy. For a system to reach that absolute zero, it would have to lose all of its energy. But to achieve that, it would have to be in contact with some other system that could accept that energy. It is possible to extract all of the internal energy from an individual atom or molecule, to bring it to its lowest internal state in which it has no more energy in vibration or rotation, but to bring a molecule to a dead stop, with no translational motion at all, in an unbounded volume, did not seem possible in the past. If the atom is confined, effectively in some sort of box or trap, then it is possible in principle to bring that atom to its lowest possible energy state in that box. That would correspond to bringing a microsystem to its absolute zero of temperature. Modern methods of experimenting with one or a few atoms at extremely low temperatures might make it possible to use light to trap an atom in a tiny "box" and remove all of its translational as well as internal energy. If this can be done, it will be a demonstration of a gap or distinction between natural scientific laws valid at the micro level and those applicable for macro systems. It might be possible to determine how small a system must be in order that, in practice, all of its energy can be removed; that is, at present, an open question. (However, that introduces a rather philosophical question as to whether "temperature" has any meaning for a single atom.) We shall return to this topic in Chapter 4, when we discuss the thermodynamics of systems that require quantum mechanics for their description.

THREE

How Did Classical Thermodynamics Come to Exist?

One particularly fascinating aspect of the history of thermodynamics is the stimulus for its origins. In the twentieth century and perhaps even a little before, we have experienced a flow of concepts from basic science into applications. This has certainly been the way nuclear energy and nuclear bombs evolved from implications in the theory of relativity. Likewise, lasers and all of the ways we use them grew out of basic quantum mechanics. Transistors emerged from basic solid-state physics. All of these examples (and others) suggest that ideas and concepts naturally and normally flow from the basic, "pure" science to "applied" science and technology, to the creation of new kinds of devices. But in reality, the flow between stimulus and application can go in both directions, and thermodynamics is an archetypal example of the applied stimulating the basic.

In sharp contrast to the "basic-to-applied" model, the origins of thermodynamics lay historically in a very practical problem. Suppose I am operating a mine and, to keep it operating, I must provide pumps to extract the water that leaks in; the water would flood the mine if it is not removed. I will fuel the pumps with coal. But I would like to know the smallest amount of coal I need to burn in order to keep the mine free enough of water to be workable. This is precisely the problem that stimulated the thinking that led to the science of thermodynamics. In the late seventeenth century, the tin mines of Cornwall in southern England were typical of those that needed almost constant pumping to keep the water levels low enough to allow miners to dig ore. Here we see how the challenge of a most practical and very specific question has led to one of the most general and fundamental of all sciences—the basic arising from the applied. Even at this early stage of our development, we get an insight into a very general characteristic of science: virtually any question we might ask about anything in the natural world may lead to new levels of understanding into how nature functions and how we can make use of that new understanding. When a new idea is conceived, or a new kind of question is raised, we have no conception of its future implications and applications.

The history of the concepts and, eventually, the laws of thermodynamics can be said to have begun in England with Francis Bacon (1561–1626), then in France with René Descartes (1596–1650), again in England with Robert Boyle (1627–1691), or even earlier in Italy with Galileo Galilei (1564–1642). These were among the early investigators who, independently, identified heat as a physical property, specifically one they associated with motion of the atoms of which they assumed matter is composed—even though nobody, at that time, had any idea what those atoms were or what any of their properties would be. Their contributions to this topic, however, were qualitative concepts; the formulation of a more precise treatment of heat, and the connection between heat and work, came a little later, through the thoughts and experiments of several key figures and through a fascinating period of controversy.

One very important step in that development came from the study of the relation between heat and the behavior of gases. Likewise, some crucial advances led to the invention and development of the steam engine. The French scientific instrument inventor Guillaume Amontons (1663–1705) showed that heating a gas increases its pressure and leads to its expansion. More specifically, he was apparently the first to recognize that all gases undergo the same expansion when they receive the same amount of heat. Furthermore, the pressure increases smoothly, in direct proportion to an increasing temperature, so that the gas pressure can be used as a measure of temperature—that is, as a thermometer. That behavior, the precise connection between temperature and pressure, later took the more exact form of a law, now known in most of the world as Gay-Lussac's law, or Charles's law in English-speaking countries. (As we saw in the previous chapter, we now incorporate the relationship of pressure, temperature, volume, and the amount of a substance into that substance's equation of state.) Amontons also recognized that the temperature of water surprisingly does not change when it boils, and that its boiling temperature is always the same, hence providing a fixed point for a scale of temperature. British scientist Edmund Halley had noticed this previously but did not realize its significance, in terms of its role in determining a fixed point on a temperature scale. Amontons also proposed a kind of steam engine, using heated and expanding air to push water to one side in a hollow wheel, creating an imbalance and hence a rotation. However, the first practical engine actually built and used was that of Thomas Newcomen (1663–1729) in England, in 1712.

Steam Engines and Efficiency

The Newcomen engine operated in a simple way (Fig. 2). A wood fire supplied heat to convert water to steam in a large cylindrical boiler equipped with a piston or in an external boiler from which steam flowed into the cylinder with its moveable piston. The steam pushed the piston, which, in turn, moved the arm of a pump, which would lift water, typically out of a flooded mine. The hot cylinder would be cooled with water, enough to condense the steam, and the piston would return to its position near the bottom of the cylinder. Then the whole cycle would start again, with heating the cylinder, and repeat itself as long as its work was needed. Each stroke of the piston required heating and

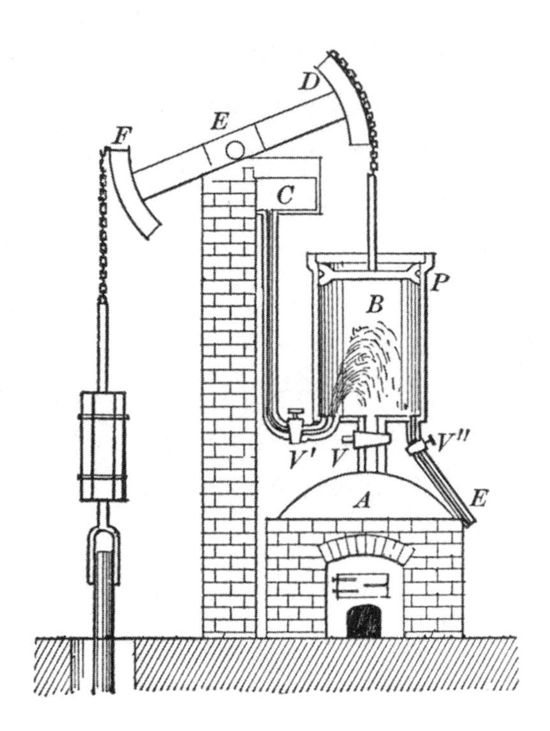

Figure 2. Schematic diagram of a Newcomen engine. Steam generated in the cylinder pushes the piston up, which in turn moves the lever arm at the top. Then water cools the steam and cylinder so the piston moves back down, pulling the lever arm down. (Courtesy of Wikimedia)

then cooling the entire cylinder. The first installed Newcomen engine drove pumps that drew water from coal mines on the estate of Lord Dudley, and soon these engines became the most widely used means to pump water from mines in England, Scotland, Sweden, and central Europe. Occasionally, others tried to introduce different kinds of heat engines, but at the time none could compete with Newcomen's. One of his, called "Fairbottom Bobs," is preserved at the Henry Ford Museum in Dearborn,

Figure 3. The Newcomen engine "Fairbottom Bobs," a real, working example of this early steam engine. (Courtesy of Wikimedia)

Michigan (Fig. 3). A functioning model of the Newcomen engine still operates at the Black Country Museum in Birmingham, England.

Chronologically, now we momentarily leave steam engines and come to the introduction of the important concepts of heat capacity and latent heat. Joseph Black (1728–1799) was the key figure in clarifying these (Fig. 4). He found, for example, that when equal amounts of water at two different temperatures were brought together, they came to equilibrium at the *mean* temperature of the two. However when a pound of gold at 190 degrees

Figure 4. Joseph Black.
(Courtesy of the U.S. National Library of Medicine)

was put together with a pound of water at 50 degrees, the result was a final temperature of only 55 degrees. Hence Black concluded that water had a much greater capacity to hold heat than did gold. Thus, generalizing, we realize that the capacity to store heat is specific to each substance. It is not a universal property like the uniform expansion of gases under any given increase in temperature. Likewise, Black, like Amontons before him, recognized that although only a little heat would raise the temperature of liquid water under almost all conditions, if the water was just at the boiling point, one could add more and more heat and the temperature would not change. Instead, the water would turn to steam. Each of those forms, ice, liquid water, and steam, we denote as a *phase*. We now associate this added heat, which Black called the "latent heat," with a change of phase of the substance. Thus, we learned that putting heat into something could affect it in different ways, by raising its temperature or by changing its form from one phase to another. The temperature rises as heat is introduced when only a single phase is present, but remains constant if two phases are present (Fig. 5).

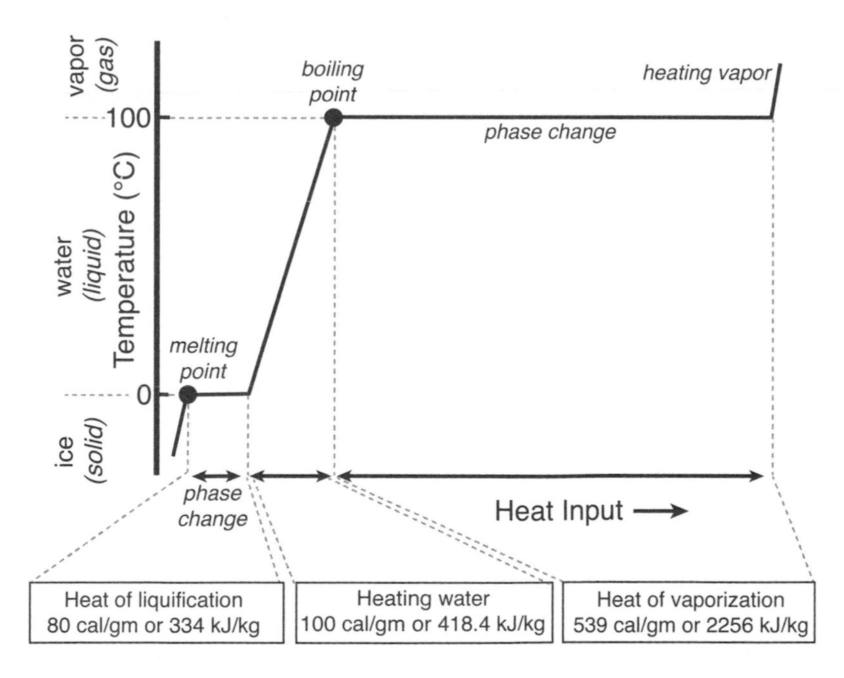

Figure 5. Latent heat diagram.
(Courtesy of Barbara Schoeberl, Animated Earth LLC)

Here we introduce a brief aside, on what constitutes a phase. There is one and only one gas phase. Likewise, there is only one ordinary liquid phase for most substances. Each structural form of a solid, however, is a phase of its own. Many solids change their form if they are subjected to high pressures. We identify each of those forms as a unique phase. Furthermore, some substances, especially those whose elements have shapes far from spherical, can exhibit both the ordinary liquid form with those elements oriented and distributed randomly (but dense, of course) and at least one other form, under suitable conditions, in which those nonspherical elements maintain some ordered orientation. We identify two or more different liquid phases for such substances. Hence *phase* is a characterization of the kind of spatial relation of the component elements of a system.

The next major advance in steam engines came half a century after Newcomen, and was a major breakthrough. James Watt (1736–1813) was a young instrument maker and scientist at Glasgow University, where Joseph Black was a professor. Watt speculated at least as early as 1760 about creating a better engine, but his breakthrough came in 1763, when he was repairing a miniature model of a Newcomen engine. He recognized that the model's cylinder cooled much more than the cylinders of full-sized counterparts, when cooling water condensed the steam. This led him to investigate just how much steam was actually generated on each cycle, compared with the amount of steam needed just to fill the volume of the expanded cylinder. Watt found that the actual amount of heat used, and hence of steam generated, for each cycle was several times what would generate just enough steam to fill the cylinder. From this, he inferred, correctly, that the excess steam was simply reheating the cylinder. He then realized that trying to improve the performance of the Newcomen engine posed a dilemma: How could one simultaneously keep the cylinder hot, so that it would not require all that reheating on each cycle, and still achieve an effective vacuum by condensing the steam to liquid water cold enough to have a low vapor pressure? At that point, his genius led him to the idea of having a second cylinder, or chamber of some sort, that would be cool enough to condense the steam, but that would only become accessible to the steam after the piston had been pushed to maximize the volume of the cylinder. That second chamber can be merely a small volume that can be opened to allow steam to enter from the main cylinder, in which that steam can be cooled and condensed to the much smaller volume of liquid water. This was the genesis of the *external condenser*, which essentially revolutionized steam engines and stimulated a whole new level of interest in improving their performance. The small external condenser could be kept cool, while the main cylinder could remain hot. The key to its use is simply the valve that stays closed except when it's time to condense the steam and reduce the volume of the main chamber. (And, of course, the condenser has a drain to remove the water that condenses there.)

Watt, who was deeply concerned with making steam engines

as economically efficient as possible, also recognized that the pressure of the steam on the piston did not have to remain at its initial high level. If the pressure was high enough initially, it would be possible to shut off the supply of steam and the steam in the cylinder would push the piston to the further limit of its stroke, with the pressure dropping all the way but still remaining high enough to keep the piston moving and expanding the chamber. At that extreme point of expansion, Watt realized that the still-hot steam would be at a pressure still barely above atmospheric, so that a valve could open to allow the steam that caused the excess pressure to condense and then flow back to the boiler. Only then would another valve open, this one to the external condenser, into which the remaining steam would flow, emptying the cylinder so the piston would return to its original position. Watt recognized that by letting the steam expand and cool almost to the surrounding temperature, one could obtain the maximum work that the steam could provide. It would be a way to obtain as much of the heat energy in the steam as practically possible (Fig. 6).

This combination of simple steps emerged from Watt's remarkable insight into how the real Newcomen engine operated, and, more important, what kinds of changes would make a steam engine much more efficient than the original model. Four related ideas—keeping the cylinder hot, using an amount of steam as close as possible to the minimum required to drive the piston through its cycle, letting it cool as it expanded, and then con-

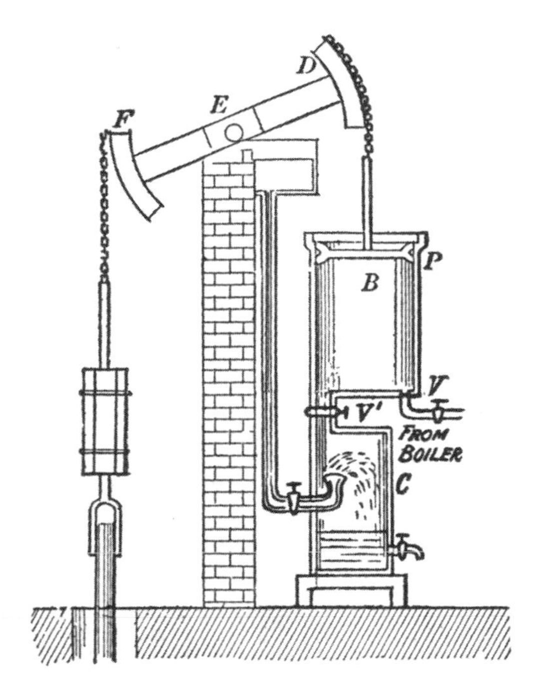

Figure 6. The steam engine of James Watt. Instead of cooling the entire large cylinder on each cycle, only the small chamber, C, receives cooling water, but C opens to the main cylinder B at the end of the expansion of chamber B, so the steam in B condenses in C, B remains hot and new water can be introduced into B and converted into steam, driving the piston P. (Courtesy of Wikimedia)

densing the remaining steam externally, outside the cylinder—enabled Watt to make an altogether far more efficient and more economical engine than its predecessors.

Another very important contribution emerged from the firm of Boulton and Watt, which made and oversaw the operations of those efficient steam engines. This was the graphical means to measure the amount of work done by the engine on each cycle, which became known as the "indicator diagram." This diagram, still constructed in its original and traditional form, represents each branch of the cycle as a curve in a graph whose axes are pressure (on the vertical axis) and volume (on the horizontal axis). John Southern, working with Boulton and Watt, invented a simple device that literally had the steam engine draw its own indicator diagram. The device consisted of a rolling table linked to the shaft holding the piston, so that the table moved back and forth as the piston moved. This motion revealed the changes in volume of the cylinder. The table held a sheet of paper. Connected to the cylinder was a pressure-sensitive arm that held a pencil, which moved across the paper, perpendicular to the back-and-forth motion of the table, with its tip on the paper, so that the pencil inscribed a curve on the paper as it and the table moved. Thus, with each cycle, the table moved and the pencil drew a closed curve of pressure as a function of piston position—that is, of the volume of the chamber. We shall address this in the next chapter, where we see that the area enclosed by the closed curve measures the work done by the engine on each cycle. Boulton and Watt incorporated these devices into their engines and thus were able to assess the performance of each one. The devices were held in locked boxes that only the employees of Boulton and Watt could open, keeping the nature of these devices secret, until someone in Australia opened one and revealed what they were and what they measured.

While Watt did make remarkable advances in the performance

of steam engines, an even better steam engine than his appeared and actually became the dominant form of engine in the Cornish mines and elsewhere. Advances by Richard Trevithick, Jonathan Hornblower, and Arthur Woolf specifically used steam at as high a temperature and pressure as possible to start pushing the piston and then continued the expansion until the steam had nearly cooled. Consistent with the later findings of Sadi Carnot, these engines proved to be even more efficient than those made by Boulton and Watt and were widely adopted.

What Is Heat?

During the years when the steam engine was being developed, the question arose regularly of just what heat is. Some of the properties of heat emerged from observations and measurements. The amount of heat required to convert a given amount of water to steam, for example, was the same as the amount one could recover by converting that amount of steam back to water at its original temperature. Melting a given sample required exactly as much heat as would be delivered when that sample again turned solid. The concept that *heat is conserved* emerged from such observations (but, as we shall see, that concept was inadequate and hence incorrect). Furthermore, people began to recognize that heat could be transferred in at least three different ways. It can go from one place to another much like light, via radiation, for example warming by the sun. It can also be transferred by the flow of a warm substance such as hot air or hot water, which we call "convection." Or it may be transferred by conduction, as it passes through a material that itself does not appear to change or move, as it does through metals, for example, but not through straw or wood. Yet these properties were not sufficient to tell what heat *is*. They only tell us something about what it *does*. The essential nature of heat remained a mystery, notwithstanding the manifold ways it was used for practical purposes.

The recognition that motion, and what we now call kinetic energy, can be converted to heat was the crucial observation from the famous cannon-boring experiment of 1798 by Count Rumford, born Benjamin Thompson in America (Fig. 7). He showed that the friction from boring out the metal from a cylinder could generate enough heat to boil the water in which the borer and cylinder were immersed. His finding laid the groundwork for the subsequent realization that mechanical work can be converted to heat and then, with heat engines, that the conversion can go in the opposite direction. Thus people came to realize that heat itself need not be conserved—but something more fundamental and general is indeed conserved, namely *energy*, as expressed in the first law of thermodynamics.

Two concepts, mutually exclusive and thus competitive, evolved regarding the nature of heat. One identified heat as the degree of motion of the (still conjectured) particles that composed materials, including solids, liquids, and gases. The other identified heat as a material substance, often as a mutually repelling matter that surrounds every material particle. The former, the kinetic

Figure 7. Benjamin Thompson, Count Rumford; engraving by T. Müller. (Courtesy of the Smithsonian Institution Libraries)

model, actually emerged first, but the material or "caloric" concept overtook the kinetic model, especially in the late eighteenth century. Two eminent and distinguished French scientists, Antoine Lavoisier and Pierre-Simon Laplace, published *Mémoir sur la Chaleur* in 1783, in which they presented both theories, without suggesting that one was superior to the other, but observing that both were consistent with the conservation of heat and with the different modes by which it can move. It was Lavoisier who introduced the term "caloric" in 1789, suggesting that perhaps he personally favored the fluid model. However, based on his friction experiments, Rumford himself rejected the idea that heat was a fluid, a concept then considered plausible by many people, obviously including Lavoisier. (Lavoisier was executed during the French Revolution; his widow, a brilliant woman herself, later became the wife of Rumford.)

According to the concepts of that time growing from the work of Joseph Black, the content of heat in any body of material is a simple function of the *heat capacity* of the material, or of the *specific heat* and the amount of material, since "specific" here refers to the heat capacity of a specific amount of the material, such as one gram. There was a belief for a time that the specific heat of any given substance is constant, but that was proven wrong. Heat capacities and specific heats definitely depend on conditions of temperature and pressure, for example. But every substance has its own specific heat, which we can think of as a property we can represent by a curve in a graph with temperature as the horizontal axis and the value of the specific heat as the quantity on the vertical axis. But we would need one such curve for every pressure, so we should think of the specific heat as represented by a surface in a space in which one axis in the bottom plane is temperature and the other is pressure, and the vertical axis is the specific heat. But constant or variable as the heat capacity might be, it was a property that could be interpreted in terms of either of the concepts of what heat *is*, motion or caloric fluid. Heat capacity could be interpreted in terms of the intensity of atomic motion or of the amount of a physical fluid. This property, important as it is for knowing how real materials behave and for explaining the conservation of heat, gives us no new insight into the nature of heat. It is simply useful as a property characterizing each substance, in terms of how it contains heat and responds to its input or outflow. Nonetheless, it will turn out to provide insight into how complex the particles are that make up each substance.

An important property of heat capacities, particularly of gases, is the way they depend on the conditions under which heat is exchanged. Specifically, we can add heat to a given volume of gas, holding the volume of the gas constant, or, alternatively, we can hold the pressure of the gas constant. If we hold the pressure constant, then the gas expands to a larger volume; if we hold the volume constant, the pressure exerted by the gas increases, as Amontons showed. We now understand why this difference exists. If the pressure is held constant, the gas, by expanding, moves the walls of its container, *doing work* in that process as it increases its temperature. If the volume is kept constant, the system increases its temperature without doing any work. We shall explore this more quantitatively shortly, but at this stage, we merely need to recognize that the extra heat required to raise the gas temperature at constant pressure is responsible for doing the work of expansion, an amount not needed for constant-volume heating. As a consequence, we distinguish heat capacities at constant volume from those at constant pressure. (These are even slightly different for solids and liquids, because they do change their volumes and densities a bit with changes in temperature—but much less than gases do.)

Another related aspect of the behavior of gases is the way they behave when they expand under different conditions. Recognizing that there are such differences was the initial step toward the discovery of the ideal gas law, the simplest of all equations of state. If a gas expands against a force, such as the pressure of the atmosphere or a piston linked to a driveshaft, and there is no heat source to maintain its temperature, that gas cools as it expands. It loses heat by doing work against that resisting force. However if a gas expands into a vacuum, its temperature remains essentially unchanged. John Dalton (1766–1844, Fig. 8), who is best remembered for his exposition of the atomic theory of matter, showed that all gases expand equally when they undergo the same increase in temperature. This finding, ironically, reinforced the belief by Dalton and many contemporaries that heat is a pervasive fluid, known as "caloric," which surrounds every atom, and that it is the increase of caloric with temperature that is responsible for the uniform expansion. This property of expansion was made more precise by Dalton's French contemporary Joseph-Louis Gay-Lussac (1778–1850, Fig. 9), who studied and compared the expansion of oxygen, nitrogen, hydrogen, carbon dioxide, and air. He showed that when the temperature of a specific volume of each of these gases increases from what we now call 0° Celsius, the freezing point of water, to 100° Celsius, the boil-

Figure 8. John Dalton. (Courtesy of the Library of Congress)

ing point, they all undergo the same increase in volume. This discovery led to the relation that the change in pressure p is directly proportional to the change in temperature, or Δp = constant × ΔT. This relation is now known in many parts of the world as Gay-Lussac's law, but because of a historical anomaly it is called Charles's law in the English-speaking world. Here we have the first precise relation that will be incorporated into the ideal gas law.

This pressure-volume relation immediately suggests that there

Figure 9. Joseph-Louis Gay-Lussac. (Courtesy of Wikimedia)

should be a limiting lowest-possible temperature at which the pressure of any gas would be zero. As we saw in Chapter 2, we now recognize this as the third law of thermodynamics, the existence—and unattainability—of a fixed, finite lower limit to what we call "temperature," which we define as "absolute zero." Dalton and others of his contemporaries attempted to determine what that lowest possible temperature is, but their estimates turned out to be far off from the actual value, which is approximately –273° Celsius, known now as 0 K, or absolute zero.

The response of gases to changes of temperature, as revealed through their heat capacities, ultimately led to a deepened understanding of the nature of gases and, eventually, of the atomic nature of matter. In particular, it was found by Gay-Lussac and then much refined by F. Delaroche and J. E. Bérard that, under ordinary room-temperature conditions, equal *volumes* of all gases at the same pressure have equal heat capacities, while the heat capacities of equal *masses* of different gases differ, one from another. Specifically, their heat capacities depend inversely on the densities of the gases. This finding eventually led to the realization, based on the atomic theory of matter, that equal volumes of gases, at the same temperature and pressure, must contain the same number of atomic particles (strictly, of molecules), but that the weights of those individual particles depend on the specific substance.

Sadi Carnot and the Origin of Thermodynamics

The origins of thermodynamics as a science—a self-contained intellectual framework to link observation, logical analysis, and a capability to describe *and predict* observable phenomena with quantitative accuracy—truly lie with the contributions of the engineer Sadi Carnot (Fig. 10). His father, Lazare Carnot, an engineer himself, was also an important figure during the mid-nineteenth century in promoting the understanding of what made machines efficient or inefficient. He was a member of a circle of French colleagues that included J. B. Biot, the mathematics professor J. N. P. Hachette, and especially Joseph Fourier,

Figure 10. Sadi Carnot as a young man, about eighteen years old. (Courtesy of Wikimedia)

who articulated general properties of heat in terms emphasizing (and distinguishing) heat capacity, heat conductivity, and heat exchange with an environment. Hence Sadi Carnot, as a young man, was aware of the growing concern about understanding heat and heat engines.

After service as a military engineer during the Napoleonic wars, in 1824 he produced his magnum opus, *Réflexions sur la puissance motrice du feu*, or *Reflections on the Motive Power of Heat.* In this volume, a crucial conceptual pillar for building ideas in thermodynamics emerged: the *ideal, reversible engine.* This is an engine with no losses of heat via friction or heat leaks to the external world, an engine that, at every instant, could be going in either direction and hence operates infinitely slowly. This concept enabled Carnot to determine the natural limit of the best possible performance that any engine can attain. "Best possible" here means *the maximum amount of useful work obtainable from any specified amount of heat.* We now call this ratio, the amount of useful work per amount of heat, the "efficiency" of a heat-driven machine, and denote it by the Greek letter η. Generalizing, we use the term to mean the amount of useful work per amount of energy put into the machine, in whatever form that energy may be. The concept of energy, however, was not yet part of the vocabulary for Sadi Carnot, and certainly the idea of using different kinds of energy to drive machines to do work lay years ahead.

As an aside, it is important to recognize that the meaning of "best possible" depends on the criterion we choose for our evaluation. Carnot used *efficiency*, which is the amount of useful work per amount of heat put in. Another, different criterion is *power*; the rate at which work is delivered; a machine optimized for efficiency will, in general, not be optimized for the power it delivers, and vice versa. One can imagine a variety of criteria of performance, each of which would have its own optimum.

As we discussed in Chapter 1, to carry out Carnot's analysis of the maximum possible efficiency of a heat-driven machine, he first invented the hypothetical engine that lent itself to just that determination of optimum performance. Then he went on to show that no other ideal engine could have an efficiency different from that of his model system, the system we now call the "Carnot cycle." This ideal engine has no friction and no exchange of heat that does not contribute to the delivery of useful work, meaning there are no heat losses through the walls of the system to the outside world. The engine operates in four steps that, when completed, bring it back to its original state.

The first step begins when the working fluid, such as air or steam, is at its highest temperature, call it T_H, and is in contact with a heat source at exactly that temperature. The fluid expands, pushing a piston to do work, and absorbing more heat from the hot heat source so the fluid's temperature remains constant. At the end of that step, the high-temperature source disconnects and the fluid expands further to reach a predetermined maximum volume, but, since it is doing work and taking in no heat, the fluid cools during this second step to a lower temperature T_L. In the third step, the now cool fluid undergoes compression, and is in contact with a cold reservoir at the same temperature as the fluid, T_L. The compression would heat the fluid but, being in contact with the cold reservoir, the temperature remains constant at its low value, T_L. Heat flows from the fluid to the cold reservoir in this step. The fourth step is another compression, but now out of contact with any external heat source or sink, so the compression process heats the fluid, until it reaches the high temperature T_H and the initial volume that began the cycle. The first and third steps are said to be "isothermal," meaning the temperature of the system remains constant; the second and fourth steps, in which the system is isolated from any outside environment, are called "adiabatic," the name for a process in which no heat passes to or from the system. (The term originates from a Greek word meaning "impassable"; it was first introduced by William Macquorn Rankine in 1866, and then used by Maxwell five years later.) As we shall see, there are many other possible cycles, but this particular one serves especially well to reveal its efficiency. The Carnot cycle can be diagrammed as a simple closed loop in a plane with x- and y-axes representing volume and pressure, respectively (Fig. 11). The top and bottom branches are carried out at constant temperature, when the system is in contact with a thermostat; the temperature changes during the other two branches when the system is isolated.

Carnot recognized that work could be generated from heat only if heat is allowed to flow from a high temperature to a lower one; no work could result if there were not two different temperatures. Using this fact together with his model of the ideal heat engine, he discovered a fundamental limit to the performance of all heat engines. He realized that if an amount of heat we can call Q_H is taken in at the high temperature T_H, and an amount of heat Q_L is deposited at the low temperature T_L, then the amount of work W would be equal to the difference between these two, $Q_H - Q_L$. This is because, with this ideal engine, there is no other place the heat can go. Consequently the efficiency, work per unit of heat taken in, is $\eta = W/Q_H$, or $(Q_H - Q_L)/Q_H$.

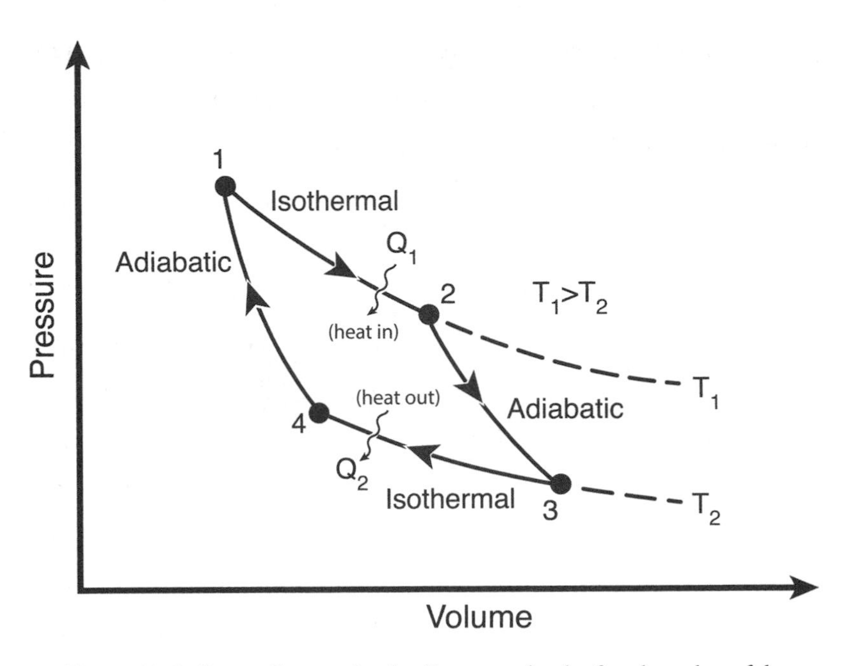

Figure 11. Indicator diagram for the Carnot cycle; the four branches of the cycle operate successively between the higher temperature T_1 and the lower temperature T_2. The pressure changes continuously with the volume during each of these steps. (Courtesy of Barbara Schoeberl, Animated Earth LLC)

But the heat Q_H taken in is proportional to the temperature T_H; specifically, it is the heat capacity C or heat exchanged per degree of temperature, multiplied by that temperature: $Q_H = CT_H$. Likewise, the heat exchanged at the lower temperature is $Q_L = CT_L$, if we assume that the heat capacity is the same at both temperatures. Therefore the efficiency of this ideal engine depends only on the temperatures between which it operates:

$$\begin{aligned} \eta &= (T_H - T_L)/T_H \\ &= 1 - T_L/T_H. \end{aligned} \tag{1}$$

This remarkably simple expression tells us that the best possible conversion of heat into work, using the cycle devised by Carnot, depends only on the two temperatures between which the system operates. Moreover if we think of the low temperature as something close to room temperature and the high temperature as the one we can control, this expression tells us to make the high temperature T_H as high as possible if we want to make the efficiency as high as possible. This was, of course, the basis for Trevithick's efficient steam engines.

One important characteristic to recognize about Carnot's ideal engine is that in order to have no heat losses to friction, the engine must operate very, very slowly—infinitely slowly, to be precise. The engine must operate so slowly that at any instant, it is impossible to tell whether it is moving forward or backward. We call such an engine *reversible*. Yes, it is unrealistic to think of using an ideal, reversible engine to carry out any real process, yet the concept serves as a limit, however unattainable, to real processes. It enabled Carnot to find that ultimate limit of efficiency for his engine. But it was more: introducing the idea of a reversible engine as the hypothetical limit of real engines was one of the novel concepts that Carnot introduced to revolutionize the way we use thermodynamics.

Then Carnot took one more major step, often overlooked in discussions of his contributions. He began with the realization that it is impossible to create a perpetual motion machine. With this as a basic premise, he showed that all ideal machines must have that same limit of efficiency. The reasoning is remarkably straightforward. Imagine an ideal engine of any kind operating between the same two temperatures as the Carnot engine, but suppose it has an efficiency greater than that of the Carnot engine. Then it can run forward, driving the Carnot engine *backward*, so that it delivers heat Q_H to the high-temperature reservoir, having extracted it from the cold reservoir. Then that heat can be used to drive the more efficient engine, which in turn pushes the Carnot engine to raise another Q_H to the hot reservoir. This means that the two engines could go on forever, together making a perpetual motion machine, taking heat from the cold reservoir at T_L and raising it to the hot reservoir at T_H. Carnot recognized that this was completely inconsistent with all our experience and therefore must be impossible. Hence the other ideal engine can be no more efficient than the Carnot cycle. The same argument tells us that the other ideal engine cannot be less efficient than Carnot's, because if it were, we could drive it backward with the Carnot engine and, in that way, raise heat from a low temperature to a high temperature without limit. Hence all ideal heat engines must have the same limiting efficiency, given by equation (1) above. This reasoning, with its amazing generality, is the culmination of Carnot's contribution and is effectively the origin of thermodynamics as a precise science.

Despite its deep innovation and key role in shaping the subject, Carnot's work was virtually unnoticed for a decade. Its first real recognition came in 1834, when Émile Clapeyron published a paper utilizing Carnot's concepts, particularly by representing Carnot's hypothetical engine via a form of its indicator diagram, the plot of the cycle of the engine as a function of the pressure and volume for each of the four steps.

What Are Heat and Energy?

The nature of heat was a central issue in the evolution of the concepts that became thermodynamics. That heat can be carried by motion of a warm fluid, whether liquid or gas, was long recognized, especially through the work of Count Rumford at the end of the eighteenth century. Recognized as well was the capability of some materials, notably metals, to transmit heat efficiently, while other materials, such as straw, could not. Another significant step came with the realization that heat can also be transmitted as radiation, which is clearly different from conduction through metals or from convection, the way fluids flow to transmit heat. It was a major step to recognize that the radiation from the sun that carries heat obeys the same rules of behavior as visible light, and that, consequently, radiant heat must be essentially the same as light, a wave phenomenon. This identification was one of the major contributions of the French physicist and mathematician A. M. Ampére (1775–1836, Fig. 12). He showed that radiant heat differed from light only in the frequencies of the two kinds of radiation. Thus, people came to recog-

Figure 12. André-Marie Ampère.
(Courtesy of the Smithsonian Institution Libraries)

nize that heat can pass spontaneously in different forms from one object to another—and that this passage always occurs from a warmer body to a colder one, never in the reverse direction.

Some insight into the capacity to do work came with the identification by G. W. Leibnitz (Fig. 13) of the product of mass times the square of velocity, mv^2, as what he and others called *vis viva*, his measure of force. But not long after, both G. G. de Coriolis and J. V. Poncelet showed that it is $mv^2/2$ that is the relevant quantity. Meanwhile, others, Descartes and the "Cartesians" (in opposition to the "Leibnitzians"), argued that mass times velocity, mv, should be the measure of capability to do work

Figure 13. Gottfried Wilhelm Leibnitz.
(Courtesy of the Herzog Anton Ulrich-Museum)

and that, in effect, this quantity is conserved. At that time, the first half of the nineteenth century, it seemed impossible that *both* could be valid and useful concepts. Today, we recognize $mv^2/2$ as the kinetic energy of an object and mv as its momentum, both of them valid, useful characteristics of a body's motion, and have forgotten altogether that at one time people believed only one or the other could be a valid "measure of motion."

The fundamental question of what heat really is remained a long-standing dispute over many years. There were those two opposing views: one that originated in the eighteenth century said that heat is the motion of the particles that make up matter;

the other held that heat is a fluid, called "caloric," something incompressible and indestructible. Of course the former, the association of heat with kinetic energy, won out, but we still use some of the language of the fluid model when we talk of how "heat flows from hot to cold." An early demonstration by Sir Humphry Davy, that rubbing two pieces of ice against each other could melt them, was invoked to support the kinetic theory, but the controversy continued well after that, at least partially because there were other interpretations of the Davy experiment that suggested it was not necessarily friction that melted the ice.

Thus, bit by bit, the different forms that heat transfer can take, and the interconvertibility of heat and mechanical work, became an acceptable, unifying set of ideas that laid the foundation for the emergence of the concept of *energy*.

That concept arose, even before it received its name, from an insight of J. R. Mayer in Heilbronn, Germany, in 1842, when he realized that energy is a conserved quantity. Mayer referred to the "indestructability of force," rather than, with all its possible transformations, what we now call energy, but the concept was indeed the same. One of the keys to his insight came from the recognition of the difference between the heat capacities of a gas held at constant volume and that at constant pressure. The gas at constant volume simply absorbs any heat put into it, so all that heat goes to increasing the temperature of the gas. The gas at constant pressure expands when heat is put into it, so that it not only experiences an increase in temperature but also does work in expanding, for example by pushing a piston. The result is that the heat capacity of the gas at constant pressure is larger, by the amount required to do the work of expanding, than that at constant volume. That difference is essentially a simple universal constant added to the constant-volume specific heat, the heat capacity per unit of gas, typically one mole. Thus if C_v is the specific heat of a gas at constant volume and C_p is its specific heat at constant pressure, then $C_p = C_v + R$ (where, as previously, R is the "gas constant"), which we realize is the energy required to expand a gas when it absorbs enough heat to increase its temperature by 1° at constant pressure. This was a key step in the recognition of work and heat as forms of something common to both of them, and then to radiation and other manifestations of what we now call energy. Mayer also demonstrated in 1843 that he could warm water by agitating it, supporting the kinetic model for heat.

Another person who played a key early role in developing the concept of energy and its conservation was James Prescott Joule (Fig. 14), professor of physics at the University of Manchester, one of the great centers of science and engineering throughout the nineteenth century. It was Joule who showed, in 1843, precisely how electric currents generate heat. By using falling weights to operate the electric generators that produced those currents, he was able to determine the quantitative equivalence between mechanical work and heat. He established this equivalence in other ways, including demonstrating the heating that can be achieved when a gas is compressed adiabatically, that is,

Figure 14. James Prescott Joule. (Courtesy of Wikimedia)

in a manner allowing no exchange of heat with the environment during the compression process. He also confirmed that if a gas, air in his experiments, expands into a vacuum, it does no work and no heat is lost. In 1847 he also showed, like Mayer, that mechanical work could be converted directly into heat, such as by simply stirring or shaking a liquid. This was strong supporting evidence for the idea that heat is motion of the elementary particles composing a system, and thus a blow to the idea that it is a fluid called caloric. His work stirred the attention of at least two important scientific figures, Hermann von Helmholtz in Germany and William Thomson (Lord Kelvin) in Britain; this happened at the time Joule described his latest, most improved device for measuring the mechanical equivalent of heat at a meeting of the British Association in Oxford in 1847. It was then that people first began to accept the conservation of energy, a concept very well articulated by Joule and strongly accepted by Kelvin. It was a major step, to recognize that something more general than heat must be conserved, because different forms of that more general property can be interconverted.

William Thomson, who became Lord Kelvin of Largs in 1892 (Fig. 15), and his older brother James played very important roles in the next steps in developing the science of energy. Working first in Glasgow, a center at least as important for the field as Manchester, they focused first on the efficiency of engines of various kinds. Both steam and heated air were the driving materials in the heat engines of that time, which by midcentury had replaced water-powered pumps. William devised a temperature scale, taking the freezing point and boiling point of water as the fixed points, with regular steps or degrees between them. His scale was essentially identical to the one developed by Anders Celsius in 1742, a century earlier. Thomson introduced the term "thermodynamic" in a crucial paper of 1849, in which he analyzed the work of Sadi Carnot—having finally found and read Carnot's nearly forgotten book. He connected that analysis to the experiments by Clapeyron and by Victor Regnault (Fig. 16),

Figure 15. William Thomson, Baron Kelvin.
(Courtesy of the Smithsonian Institution Libraries)

measuring the ratio of the work done to compress a gas to the heat released by that compression. Specifically, these experiments were meant to determine a quantity fundamental to quantification of Carnot's analysis, the maximum amount of work that could be done by a temperature change of 1°.

Thomson's analysis of those experiments effectively demonstrated that there should be an absolute scale of temperature, with

Figure 16. Victor Regnault.
(Courtesy of the Smithsonian Institution Libraries)

a lowest value at –273° C. This was a much more precise determination than the inference made by Amontons a century earlier, that air would lose all its elasticity at a temperature of about –270° C. Although Thomson recognized both that there was a natural "fixed point" and what its value was, he did not himself introduce the Kelvin temperature scale. With its zero point at –273° C, now recognized as the lowest temperature attainable in principle, the Kelvin scale has a true natural fixed point, so it is an absolute scale of temperature. This is in contrast to the arbitrary choices of the Celsius scale or the Fahrenheit scale, which place the freezing point of water at 0° C and 32° F, respectively.

The emergence of *energy* as an encompassing concept was largely the result of work by Kelvin and especially by his German contemporary, Hermann von Helmholtz. In 1847, Helmholtz published a paper with a title rather freely translated as "On the Conservation of Energy," which uses Joule's results to infer precisely that idea. He made his inference from two premises: one, that there are no perpetual motion machines, and the other, that all interactions involve the attractive and repulsive forces between the elementary particles that matter comprises. Incidentally, he used the then established expression *vis viva*, and also "Kraft," strictly translated as "force," rather than "energy" for the conserved quantity. The distinction between force and energy was not yet clear at that time. Kelvin, then still William Thomson, discovered Helmholtz's work in translation in 1852 and used it to unify the mechanical, thermal, and electrical manifestations of energy, effectively establishing the conservation of energy as a fundamental principle in itself. In his own paper of 1852 built on Helmholtz's work, he stated explicitly that *energy* is conserved, using that word. His Scottish contemporary Macquorn Rankine, another important figure in the development of thermodynamics, was the person who introduced the explicit expression "the law of the conservation of energy." Rankine was also the first to draw a distinction between potential energy and what he called "actual energy" (which we call "kinetic energy"). It was with his usage that the word "energy" became fully established over the various previous wordings. It was quickly adopted by the Thomson brothers, and by the Germans including Helmholtz and another important figure, Rudolf Clausius. It was Rankine who, in a paper in 1857 and a book in 1859, first used the term "principles of thermodynamics" and cited "the first and second laws of thermodynamics." Although that citation incorporated work from an early paper by Clausius, at that time the second law did not yet have the precision it was given by Clausius in the following years.

It was Rudolf Clausius, a German physicist, who first enunciated the second law of thermodynamics. He was a professor in Zürich when, in 1854, he published his first statement of the law: that heat can never flow spontaneously from a colder to a warmer body, and, by implication, that work must be done to make that happen. In 1865, he introduced explicitly the concept of "entropy," as well as the word, giving it a precise mathematical definition. It was then that he expressed the first and second laws in essentially the words we use today: The energy of the universe is constant (the first law), and the entropy of the universe tends to a maximum (the second law). Until his statement, the concept of heat flowing from a cold body to a warmer one was apparently consistent with the conservation of energy, yet it is obvious from all of our experience that this doesn't happen. Hence it was necessary to enunciate a second fundamental principle, which is precisely the second law, to be an integral part of the basis of the new science of thermodynamics.

The precise statement of that law, mathematically, is in terms of an *inequality*, rather than as an equation. If a small amount of heat, δQ, is added to a body whose temperature is T, then the entropy δS of that body increases, satisfying the *inequality* relation $\delta S \geq \delta Q/T$. The equality constitutes a lower bound for the change of entropy. The concept of entropy, also called "disgregation" by Clausius, was, at that time, still imprecise by today's standards, but it was recognized to be a measure of disorder, in some sense. Only later, when thermodynamics was linked with the microscopically based statistical mechanics, largely by Ludwig Boltzmann of Germany, and then more precisely and elaborately by an American, J. Willard Gibbs (Fig. 17), did entropy acquire a truly precise, quantitative meaning, which we shall examine briefly later. Here, we need only recognize that (a) entropy is a quantity measured in units of heat per degree of temperature, such as calories per kelvin, and (b) that there is a lower bound to the amount it increases if a system acquires heat. In fact, of course, entropy can increase even if no heat is injected into a system. A gas, initially confined to a small volume, expands spontaneously into an adjacent, previously evacuated larger chamber, with no input of heat. A drop of ink that falls into a glass of water eventually diffuses to become uniformly distributed in the water, with no input of heat. A raw egg, dropped, breaks into an irreversible state. All of these examples illustrate the natural increase of entropy in the universe. For all three, δQ is zero, so we can say that $\delta S > 0$ for them. If a system is cooled by some refrigeration process (so work is done), then δQ is negative, and

Figure 17. Josiah Willard Gibbs. (Courtesy of the Beinecke Rare Book and Manuscript Library, Yale University)

the entropy change may also be negative, but if it is, it must be closer to zero than $\delta Q/T$. The equality sign can be appropriate only for ideal, reversible processes, never for real processes that involve heat leaks, friction, or other "imperfections."

Entropy and Statistics: Macro and Micro Perspectives

In the mid-nineteenth century, a serious issue developed from the apparent incompatibility of Newtonian mechanics and the evolving second law of thermodynamics. Newton's laws of motion implied that all mechanical processes must be reversible, so that any process that can move "forward" can equally well move "backward," so that both directions, being compatible with the laws of motion, should be equally likely. On the other hand, the second law of thermodynamics overtly gave a specific direction in time to the evolution of the systems it describes. The laws of motion seem to imply that if a gas, a collection of molecules, in a volume V, is allowed to expand into a larger volume, say 2V, that it is perfectly reasonable to expect that, at some time, those molecules would all return to the original volume V. But the second law of thermodynamics says that we'll never see it happen.

Here we encounter a fundamental problem of reconciling two apparently valid but incompatible descriptions of how matter behaves. One, based on the laws and equations of motion of individual bodies, looks at matter from an individual, microscopic point of view. The other, based on thermodynamics, looks at matter in the larger sense, from a macroscopic point of view. In some ways, the problem of reconciling these two viewpoints continues to arise in the natural sciences. This problem led to two conflicting views of the second law during the late nineteenth century: is the second law *strictly* true, with entropy always increasing, or is it valid only in a probabilistic sense? This could be seen as a question of the compatibility of Newtonian mechanics, for which all motions are equally valid with time running forward or backward, and the second law, with its inherent direction of time. However, the problem of reconciling the second law with Newtonian mechanics did achieve a satisfactory resolution through the recognition of the importance of statistics, and the creation of what we now call statistical mechanics, or statistical thermodynamics. And of course, the validity of the second law does rest in the nature of statistics and statistically governed behavior. In particular, statistics tells us that the more elements there are that constitute the system, the less likely are recognizable fluctuations from the most probable microstates. The second law is, strictly, valid only in a probabilistic sense, but for systems of macroscopic size, the probabilities weigh so heavily in one time direction that there is essentially no likelihood that we would ever observe a macroscopic system evolving spontaneously from a higher to a lower entropy.

We shall describe how this approach emerged historically, but for the moment, it suffices to say that, yes, the gas molecules could perfectly well go back from occupying volume 2V to being entirely in the initial volume V, and the laws of mechanics virtually assure that at some time, that would happen. But we can examine this in a bit more detail. Having all the molecules return to V is a very, very unlikely event for any large assembly of molecules. For two or ten molecules, it is quite likely that we could, in some reasonably short time, observe all of them returning from 2V to V, however briefly. But if we were dealing with 1,000 molecules, such a return would be quite improbable, and if we consider a sample, for example, of all the air molecules in a volume of 1 cubic meter, and allowing them to expand to 2 cubic meters, the probability of their returning to the original 1 cubic meter is so infinitesimal that we could expect to wait many, many times the age of the universe before we would see that happen. And even then, that condition would persist for only an infinitesimal time. We'd have to be unbelievably lucky to catch it in our observations. To get a feel for the number of molecules in a typical macroscopic system, recall that we typically count atoms and molecules on this scale using units we call *moles*. The number of particles in a mole is the number of atomic mass units (with the mass of a hydrogen atom equal to 1) in a gram. That number, Avogadro's number, is approximately 6×10^{23}. Hence, for example, one mole of carbon atoms, atomic weight 12 atomic mass units, is just 12 grams, an amount you can easily hold in your open hand.

In other words, the macroscopic approach of thermodynamics has at its roots the assumption that the atoms and molecules that compose matter obey statistics of large numbers that describe random behavior. Macroscopic science depends inherently on the properties of systems of very large numbers of component particles. That means that what we observe of a system at any time is determined by what is the most probable way that system's constituent particles can be, given how the system was prepared. The key here is that the more constituents that the system contains, the more the *most probable* condition and tiny fluctuations from that condition dominate over all others. Having essentially uniform density of air throughout the 2-cubic-meter volume can be achieved in so many more ways than having measurable fluctuations that deviate from uniform density that we simply never see those potentially measurable fluctuations. Yes, Newtonian mechanics and the requirement that the molecules move randomly does assure that those fluctuations can, and even will, occur—in principle. But observably large fluctuations are so rare for macroscopic systems that we would never observe them. We can't wait that long, and they wouldn't last long enough to be clearly distinguishable, either.

There is another aspect to the reconciliation of the micro and macro approaches to describing the world. Newtonian mechanics is indeed formally reversible, so that if we specify precisely the positions and velocities of all the particles of a system at some instant, the system can go forward, and at any later time, if we reverse the velocities, the system will return to its initial state. In reality, however, we can never describe those initial positions and velocities *precisely*, meaning that we can specify them only to some number of significant figures. Beyond that limit, we have no knowledge. But then, as the real system evolves, with the particles interacting, colliding and bouncing off one another, our knowledge of the positions and velocities becomes less and less precise. Each collision increases our uncertainty of the microstate of the system, ultimately but in a finite time, to a point that stopping the system and reversing its direction would have at most an infinitesimal probability of returning to that initial state, even to just the imperfectly specified state that we knew it to have at the initial instant. This decrease in our knowledge of the system reveals another aspect of entropy, namely its connection with information—or lack thereof.

Let us look a bit deeper into the precise, quantitative concept of entropy. As we discussed in Chapter 2, we again need to distinguish *macrostates* from *microstates*. A macrostate of a system is one describable by thermodynamic variables, such as temperature and pressure, if it is in equilibrium. If it is not in true equilibrium, then perhaps it may be described by those thermodynamic variables plus additional variables specifying the bulk flows and any other nonstationary properties, such as the mean velocity of the water in a river. We can, after all, speak of a fluid that cools as it flows through a pipe, assigning a temperature to all the positions along the pipe; such a system is not in equilibrium but is still describable by thermodynamic variables, augmented with others such as the flow velocity. In contrast, a microstate is one in which the positions and velocities of all the elementary particles that make up the system are specified. Obviously we would not care to describe, for example, the air in a room, even at equilibrium, in terms of its microstate. The important point here is that, for a macroscopic system, the number of microstates that are consistent with its macrostate is vast, so enormous that we would not wish to deal with them. However, the number of microstates corresponding to any chosen macrostate is important; a macrostate that can be achieved by some moderately large number of microstates is still different from one that can be achieved by, say, an even much larger number of microstates. In the former, if it does not have too many elementary components, we may be able to observe fluctuations from the most probable; in the latter, the probabilities of observable fluctuations are simply far too small for us ever to see them. The extent of the fluctuations in a random system of N particles, such as the air in a room, is proportional to $\sqrt{N}$, which grows much more slowly than N itself. Hence the larger the system, the narrower is the range of fluctuations and the more difficult they are to observe. (One interesting challenge, with current capabilities to observe small systems, is determining the approximate maximum size system for which we might observe those fluctuations.)

To deal with this situation of large numbers, we use not the number of microstates that can yield a given macrostate, but rather the logarithm of that number, the exponent, normally taken to be the exponent of the natural number *e*, or Euler's number, which is approximately 2.718—the power to which e must be taken to reach the number of microstates. (The number *e* is the limit of the infinite sum $1 + 1/1 + 1/(1 \cdot 2) + 1/(1 \cdot 2 \cdot 3) + \ldots$) If we call the number of microstates corresponding to a chosen macrostate N, then the quantity we use is called the natural logarithm of N, written ln(N). Hence $\exp[\ln(N)] = e^{\ln(N)} = N$. Thus we use ln(N) as the basis of the microscopically based definition of the entropy of that state. Readers may be more familiar with logarithms based on powers of 10, which we write as log(N). Thus log(100) is 2, since $10^2 = 100$.

Logarithms are pure, dimensionless numbers. In normal usage, we give the entropy its numerical scale and dimension of energy per degree of temperature, by writing the entropy S as either $k \ln(N)$, counting individual atoms, or $R \ln(n)$, counting a fixed large number of atoms at a time, specifically Avogadro's number. Here k is the Boltzmann constant, approximately 1.38×10^{-23} joules per kelvin, or 1.38×10^{-16} ergs per kelvin, and R is the gas constant, just the Boltzmann constant multiplied by approximately 6×10^{23}, the number of particles in a mole (the number of individual hydrogen atoms that would weigh 1 gram). Avogadro's number is the conversion factor that tells us the number of atomic weight units in 1 gram (since the hydrogen atom weighs essentially 1 atomic weight unit). Working with macroscopic systems, we typically count atoms or molecules by the number of moles they constitute.

Now we can tie the statistical concept of entropy to the second law: systems changing their state spontaneously always move from a state of lower entropy to a state of higher entropy—in other words, they always move to a macrostate with a larger number of microstates. The more ways a macrostate can be achieved in terms of microstates, the more probable it is that it will be found, and the greater its entropy will be. The precise relation, first enunciated by Boltzmann in 1877, for the entropy S of a macroscopic system is simply $S = k \ln W$, where W is the number of microstates that correspond to the macrostate of interest.

There is one other powerful kind of tool that thermodynamics provides. We previously mentioned *thermodynamic potentials*, the natural limits of performance of processes. These are variables that balance the tendency for systems to go to states of low energy (and typically low entropy) and to states of high entropy (and likewise, typically high energy). The specific form of each depends on the conditions under which it applies. For example, if the temperature and pressure of the process are the controlled variables under which a system is evolving, the appropriate thermodynamic potential is called the Gibbs free energy, which takes the form $G = E - TS$, so, since it is useful under conditions of constant temperature T, its changes for such conditions are expressed as $\Delta G = \Delta E - T\Delta S$.

An application of this particular potential gives us the ratio of the amounts of two interconvertible forms of matter, such as two chemical forms of a substance, say A and B. We call that ratio, which depends on the temperature and pressure, an *equilibrium constant*, K_{eq}. Explicitly, the ratio of the amounts [A] and [B], when the two are present in equilibrium, is given by the relation $K_{eq} = [A]/[B] = \exp(\Delta G/kT)$, where $\Delta G = G(A) - G(B)$ and the G's, the Gibbs free energies per molecule of A and of B are in atomic-level units. (If we use molar-level units for the amounts of A and B, we must replace the Boltzmann constant k with the gas constant R.)

It is sometimes useful to express the thermodynamic potentials in terms of their values per particle, instead of in terms of the total amounts of the substances; if we count by atoms or molecules, rather than by moles, we write $G = N\mu$, or, for a change, $\Delta G = N\,\Delta\mu$. The free energy per atom, μ, is called the *chemical potential.* The thermodynamic potentials, which include energy itself, the Gibbs free energy and other parallel quantities, are powerful analytic tools for evaluating the real and potential performance of real engines and processes. They enable precise comparisons of actual and ideal operations, in order to help identify steps and stages with the greatest opportunity for technological improvement. We shall return to the Gibbs free energy and the expression for it in the next chapter.

Now we can examine how statistics came to play a central role in the natural sciences, and specifically how it provided a link from the macroscopic approach of thermodynamics and the microscopic approach of mechanics, whether Newtonian or the later quantum variety—which brings us to our next chapter.

FOUR

How Do We Use (and Might We Use) Thermodynamics?

Thermodynamics is, as we have seen, a basic science stimulated by the very practical issue of making steam engines operate efficiently. Carnot realized from the expression he derived for efficiency that it is desirable to make the temperature of the high-temperature heat as high as possible, and the temperature of the cold reservoir as low as possible. This naturally played a role in how steam engines were designed, and continues to be a basic guideline. But thermodynamics has given us many other insights into ways we can improve technologies that derive useful work from heat.

A class of very straightforward applications of thermodynamics consists of the cooling processes we use, the most obvious being refrigeration and air conditioning. These achieve the desired lowering of temperature by evaporating a volatile substance; when a condensed liquid converts to vapor, this process requires the evaporating material to absorb energy from its surroundings and hence to cool those surroundings. But we expect refrigerators to operate for long periods, not just once, so we must re-condense that material back to liquid, a process that requires work. Therefore, refrigerators operate by going through cycles much like the one we saw in the indicator diagram for the Carnot cycle in the previous chapter. Instead of going clockwise, however, the direction that generates work by extracting heat from a hot reservoir and depositing the energy that wasn't converted to work into the cold reservoir, a refrigerator runs counterclockwise around its cycle. The refrigerator removes heat from the cold reservoir, first by evaporation, and then undergoing an adiabatic expansion, a process that cools the vapor but allows no energy exchange with the surroundings during the process; it then deposits heat at the higher temperature when the vapor is compressed and liquefied, and of course does work in that step—rather than producing work. The area within the cycle is precisely the work we must supply to go around the cycle one time—if the system has no friction or heat leaks, real imperfections that of course require some extra work. Because the cycle follows a counterclockwise path, the area within the cycle has a *negative sign*, indicating that the area gives us the amount of work we must supply for each cycle—in contrast to the clockwise paths we saw previously, whose enclosed area carried a positive sign, revealing the amount of work provided to us on each cycle. Naturally there must be some way for the heat extracted in the cooling process to leave; we normally dissipate that heat into the surroundings, into the room where the refrigerator operates, then into the air outside. Otherwise the refrigerator would heat its immediate surroundings!

We can think of the refrigeration process as a means to maintain a low entropy in whatever it is that we are cooling. So long as that food or whatever it is remains at a temperature below the room temperature of the surroundings, its entropy is lower than what it would be if it were allowed to come to thermal equilibrium at room temperature.

Of course one of the most widely used applications of thermodynamics, in the sense of converting one form of energy into another, is the generation of electricity, the conversion of any of a variety of forms of energy into that one specific form. We typically use energy as heat, typically in the form of hot steam, as the immediate form of energy to drive generators, so we convert that heat into mechanical energy, which, in turn, we use to produce electrical energy. But the heat comes, in most of our power stations, from combustion of coal, oil, or gas, which means the primary source from which we eventually obtain electric energy is actually chemical energy, stored in the fuel. We are learning, however, to convert the kinetic energy of wind and even of ocean water into the mechanical energy of generators, and to convert radiation energy in sunlight directly into electrical energy. We even know now how to convert energy stored in atomic nuclei into heat in a controlled way, then into mechanical energy and ultimately into electrical energy—it's of course called "nuclear energy."

One approach whose application continues to grow is called "combined heat and power" or CHP. It also goes by the name "cogeneration." This makes use of the obvious fact that heat itself has many uses, so that the heat produced when we generate electricity can be put to use, rather than just discarded into the environment. After the very hot steam has given energy to the mechanical generator, it is still hot, albeit not as hot as it was. The energy remaining in that steam can be used, for example, to warm buildings. This is probably the most widely applied form of combined heat and power. More than 250 universities, including Yale, now use CHP to generate their electricity and heat their structures. Typical overall efficiencies of separate generating and heating systems are about 45%, while combined heat and power systems can have overall efficiencies as high as 80%. (Here, efficiency is the amount of energy put to deliberate use, whether as heat or electricity, per energy input.)

The substantive information contained in thermodynamics is a constant guide to how to get the maximum benefit from whatever energy conversion process we are using, whether it is conversion of heat to mechanical energy that we use directly, or then transform that mechanical energy to electrical energy, or convert chemical energy to heat or directly to electrical energy, as in a battery. We learned from Carnot that in converting heat to mechanical energy, the high-temperature step of a cycle should be carried out at as high a temperature as possible. The simple comparison of real and ideal machines tells us to minimize any subsidiary processes, such as friction and heat leaks, so we learn to make efficient lubricants and thermal insulators. (Sometimes we inadvertently create new problems when we try to minimize those losses. An example is the wide use people made of asbestos as a heat insulator—for which it is indeed very good. However it was some time before the harmful health effects of inhaling asbestos were recognized, and when that happened, other materials replaced asbestos in many applications.)

Another closely related application from thermodynamics is in the design of what is called an open-cycle turbine. In such a device, a compressor takes in air mixed with fuel, typically natural gas (methane, specifically), and the mixture burns, generating heat and warming the gas mixture, thereby increasing its pressure. At that high pressure, the hot gas drives the turbine wheel of an electric generator. The more heat the hot gas contains, the more efficient is the turbine. Hence such devices are designed to capture heat left in the burned gas after it leaves the turbine; they are built to transfer heat from the gas exiting the turbine to gas that has not yet undergone combustion. In a sense, this is a sort of combined heat and power system, rather different from the one discussed above.

A growing application of improved efficiency in producing a useful form of energy is in conversion of electrical energy into visible light. Before the invention of the electric light, people produced their light directly from chemical reactions that release energy as heat and light—in flames, naturally. Candles convert chemical energy to light and heat by enabling oxygen to combine with substances that contain carbon, such as wax. Gas lamps used methane, "natural gas," as the combustible material but, like the candle, generated light from a chemical reaction that releases considerable energy.

The incandescent lamp produces light in an altogether different way, by passing electricity through a metal filament, typically tungsten, because it is a substance that, as a thin piece of wire, can withstand heat for long periods, even when raised to temperatures at which it radiates visible light—so long as no oxygen is present. The incandescent light bulb therefore has its tungsten filament sealed inside a transparent glass bulb, where there is either a vacuum or an inert gas atmosphere. This device was an enormous advance and became the principal source of "artificial light" throughout the world.

Yet more energy-efficient ways to generate visible light from electrical energy eventually appeared. One is the fluorescent lamp, in which an electrical discharge operating continuously in a gas keeps exciting a gas filling the lamp; the excited atoms or molecules collide with fluorescent material on the walls of the lamp, and transfer their excitation energy to the fluorescent material, which then emits visible light.

Another technology, now widely used, is the "light-emitting diode," or LED, a solid, typically quite small, that has a portion in which electrons are available to move if a voltage is applied to "push" them and a portion where there are empty sites where electrons can go, "fall in," and give off visible light as the way they lose what is then their excess energy. Both the fluorescent bulb and the LED produce a given amount of light from smaller amounts of input electrical energy than the incandescent lamp requires. They achieve their higher efficiency by converting the excitation energy directly into light, either from the atoms struck by electrons in the electric discharge of the fluorescent lamp or from the electrons excited by an applied voltage to move from the "donor" side to the "acceptor" side of the LED, where their excitation energy becomes light energy. In contrast, the incandescent lamp gives off visible light when the filament is heated to such a high temperature that it radiates a whole spectrum, visible light as well as infrared and other electromagnetic radiation over a broad range of wavelengths and frequencies. The fluorescent and LED lamps emit radiation only of specific frequencies, corresponding to the precise amounts of energy released when electrons lose their excess energy or the fluorescent material excited by the electrons changes from a specific high-energy state to a more stable one at lower energy. So it may be a serious challenge to create a fluorescent lamp or an LED that has some specific, desired color, particularly if that color doesn't coincide with a known emission frequency, such as the yellow sodium line so apparent in many street lights.

This brings us to an explanation that comes directly from thermodynamics, of just why hot systems emit light. Because radiation is a form of energy that most materials can absorb or emit spontaneously, if an object is not somehow isolated from its environment, it will exchange energy as electromagnetic radiation with that environment. If the object is warmer than the environment, there is a net flow of radiation out of the object; if the environment is warmer, that flow is inward. The radiation from a typical heated object consists of a broad spectrum of wavelengths, whose shape and especially whose peak intensity depend on the temperature. We refer to such an object as a "black body," in contrast to something that emits its radiation at a few specific wavelengths, such as a mercury vapor lamp. If the "black body" is fairly warm, that peak is likely to be in the infrared part of the spectrum, where the wavelengths are longer than those of visible light, or, for a barely warm object, even in the microwave region, where wavelengths can be centimeters or meters long. If the object is hot enough, however, its most intense radiation can be in the visible region of the spectrum, where the wavelengths are between 400 and 700 nanometers, from blue to red. (One nanometer is one billionth of a meter, written 10^{-9} meter.) The hotter the emitting system is, the shorter—and more energetic—is the peak intensity of the radiation it emits. A hot emitting object that looks blue or green must be hotter than one that looks red. If an object is hot enough, its peak radiation can be in the (invisible) ultraviolet region, with a wavelength shorter than 400 nanometers, or even in the X-ray wavelength region, from 0.01 to 10 nanometers.

Figure 18. Max Planck.
(Courtesy of the Smithsonian Institution Libraries)

An interesting aside is worth noting at this point. The classical theory of thermodynamics of radiation included a prediction that was obviously wrong. Specifically, that classical theory implied that the radiation from any object would become more and more intense at shorter and shorter wavelengths, so that the object would emit an infinite amount of radiation at arbitrarily short wavelengths. This was obviously wrong, but what was the explanation? It came with the work of Max Planck (Fig. 18), who proposed that energy has to come in discrete units, packages called "quanta," and that the energy in a single quantum is a fixed amount, a function of wavelength λ (or frequency ν), given by a constant, universally written as h, times the frequency, so a single quantum or minimum-size package of radiation carries an amount of energy $h\nu$. This means that radiation energy of high frequency and short wavelength—and high energy—can come only in very energetic, "big" packages. If a system has only some finite amount of energy, it can't produce a quantum that would have more than that amount. That condition thereby limits the amount of short-wavelength radiation any object, however hot, can radiate. This acts as a natural limiting factor that keeps finite the amount of radiation anything can emit.

Planck's explanation of the limitation on high-frequency radiation led to the concept of *quantization of energy*, which in turn stimulated the entire quantum theory of matter and radiation. These developments demonstrate how our understanding of nature and the world around us is always incomplete and evolving, as we recognize limitations in the explanations we have and use, and as we find new ways to explore natural phenomena. There are, even at the time of this writing, unexplained and poorly understood but well-observed phenomena that call for new, as yet uncreated science: the source of much of the gravitation around galaxies, which we call "dark matter," and the source of the accelerating expansion of the universe as revealed by the rate at which distant galaxies move away from us, which we call "dark energy."

FIVE

How Has Thermodynamics Evolved?

In the context of the steam engine, we first encountered the indicator diagram, a graph showing pressure as a function of the volume for a typical steam engine. This is a closed loop, exemplified by the indicator diagram for the idealized Carnot engine that we saw in Figure 11. There are many kinds of real engines, with many different cycle pathways; in reality, no operating engine actually follows a Carnot cycle. Traditional automobile engines, burning gasoline and air when their mixture is ignited by a spark, follow what is called an "Otto cycle." Diesel engines follow a different cycle called a "Diesel cycle." These differ simply in the particular pathway each follows. The important point for us here is to recognize something that Boulton and Watt and their engineer and later partner John Southern recognized, and that still remains an important diagnostic application of thermodynamics: that the area within the closed figure of the indicator diagram is the actual work done in a single cycle.

We see this because (a) the work done to produce a small change of volume dV against a pressure p is just the product of these two, pdV; (b) if we follow one curve of the indicator diagram as the arrows indicate, say the highest branch of the loop, by adding all the small amounts of work that are the result of going from the smallest to the largest volume on that branch of the loop—that is, adding up all the small pdV values—we would obtain the total work done when the system moves along that curve.

We call that process of adding all the tiny increments "integration," and we symbolize it as $\int$ pdV, which we call the integral. Each increment, each product, is just the area under the curve for that tiny interval dV. Hence the integral from beginning to end of that top branch is the area under that branch! That area is just the work done by the system as it expands along that top curve, as the volume increases. We can go on to the next branch, and see what work is done there. Then, when we come to the long, lower branch, a compression step, the volume diminishes with each dV so work is done on the system. The process of integration along each of the lower branches involves volume changes that reduce the size of the chamber, so there, the dV values are negative, work is done on the system, and hence the areas under the lower branches are negative. Thus, when we add the integrals, the areas beneath all four branches, those from the top two are positive, but those from the bottom two are negative, so the sum of all four is just the difference between the two sets, the area within the closed loop. This is precisely why finding the area of the loop of the indicator diagram is a straightforward way of determining the work done on each cycle by an engine, any cyclic engine.

We can construct the diagram for the ideal performance of any cyclic engine we can devise. Then we can let the machine draw its own indicator diagram, as Boulton and Watt did; we can compare it to the ideal diagram for the same ranges of volume and pressure. The comparison of the areas of the two loops immediately tells us how near the actual performance comes to that of the ideal engine.

An example is the indicator diagram for an ideal Otto cycle (Fig. 19). This cycle is actually more complex than the Carnot cycle, because it is a "four-stroke" cycle. In the Carnot cycle, one complete cycle has the piston moving out and back, in what we call two "strokes." In contrast, in the Otto cycle, the piston moves back and forth twice in each full cycle, doing useful work only in one expansion step. That is the step represented by the highest curve in the diagram, the step in which the gasoline burns in air, forcing the piston out as the gases in the cylinder expand. At the end of that stroke, the gases, still at fairly high pressure, escape through a now open valve, so the second stroke consists of a reduction of pressure while the volume remains constant. Then the piston returns, at constant pressure, to its position correspond-

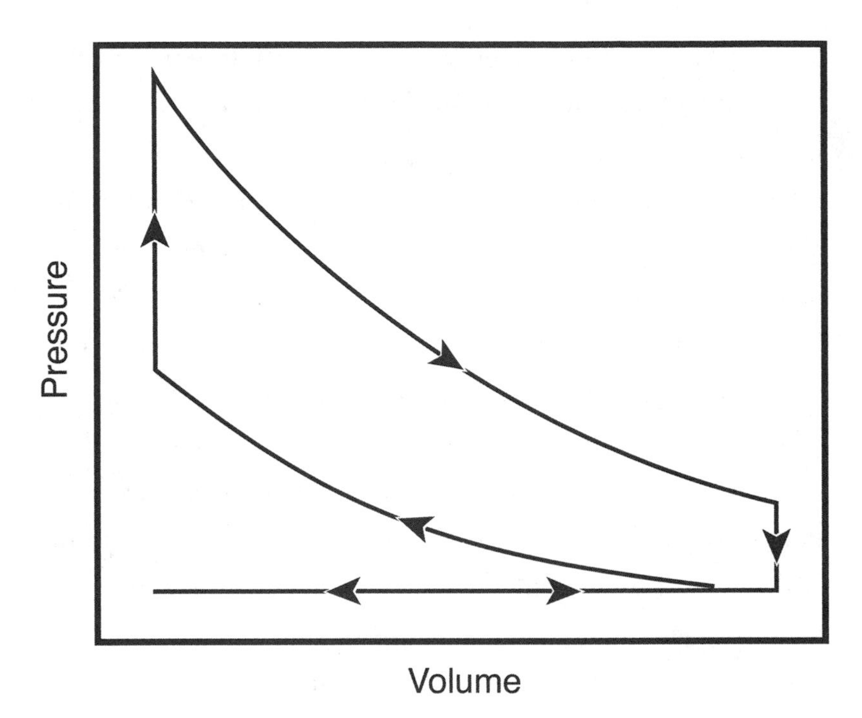

Figure 19. Indicator diagram for the ideal Otto cycle that drives the gasoline-powered automobile. Unlike the earlier diagram for the Carnot cycle, this diagram uses pressure and volume for the axes, rather than temperature and volume. On each full cycle, the system moves twice between the lowest and highest volume; in the upper loop, the cycle generates work equal to the area within that loop; the lower "back-and-forth" line is merely emptying and refilling the cylinder. (Courtesy of Barbara Schoeberl, Animated Earth LLC)

ing to the minimum volume of the cylinder (to the left in the diagram), driving the rest of the burned gases out. Next, the exhaust valve closes and the intake valve opens, fresh air enters the cylinder and the piston moves out, still at constant pressure. These two steps correspond to the back-and-forth horizontal line in the diagram. Then the piston again moves back to compress the air in the cylinder, preparing for the introduction of gasoline. That step is the one represented by the curve rising from right to left at the lower side of the loop in the diagram. At the end of that step, gasoline enters the cylinder, either by simple flow through a now open valve or through a jet. Then the spark goes off, igniting the gasoline-air mixture. The introduction of the gasoline and the firing of the spark correspond to the vertical upward line in the diagram, the step that occurs so suddenly in the ideal engine that the volume does not change. That brings the engine back to the "power stroke" from which the automobile derives its motion.

Thus a practical diagnostic of the performance of a real engine is the comparison of the area in the cycle of the ideal engine's indicator diagram with the corresponding area in the cycle of the real engine's indicator diagram. We learn from this comparison how near to the ideal limiting performance our real engine actually comes. In practice, we know the area within the cycle for the ideal engine, and we can measure the work done, per cycle, by the real engine by a variety of means, in order to make the comparison and hence evaluate the performance of our real engine. This gives us a diagnostic tool to identify potential ways to improve the performance of an engine. For example, we learn from Carnot that the highest temperature in the cycle should be as high as possible, so we design internal combustion engines to withstand very high temperatures and to lose as little of that heat through the engine's shell as possible. We could, if we find a way to devise the appropriate mechanical link, capture as much of that high-temperature heat as possible by allowing the piston to move fast while the burned fuel is at its hottest; that represents the kind of intellectual challenge that we encounter as we try to use thermodynamics to improve performance of real processes.

Linking Thermodynamics to the Microscopic Structure of Matter

A major step in the evolution of the science of thermodynamics was precisely the establishment of a bridge between the macroscopic description of matter characteristic of the traditional approach via properties of "human-size" systems and the microscopic view, rooted in mechanics, which treats matter as composed of elementary particles—atoms and molecules, typically. This was triggered in large part by the apparent incompatibility of the reversibility of Newtonian mechanical systems with the irreversibility demanded by the second law. Scientists in Germany, England, and the United States developed that bridge, called statistical thermodynamics, a name introduced by J. Willard Gibbs. The essence of this subject is the incorporation of the statistical behavior of large assemblages into the micro-oriented, mechanical description of the individual elementary constituents—atoms or molecules—as they interact with one another.

A second very important later advance that had a major influence on thermodynamics was the introduction of quantum mechanics. While it is predominantly concerned with microscopic behavior of matter and light, quantum mechanics has some very strong implications for thermodynamics, via the allowable statistical properties of the microscopic particles, especially at low temperatures. We shall examine those implications in this chapter.

James Clerk Maxwell (Fig. 20) was a believer in the idea that matter was made of small particles that could move about and interact with one another. In gases, these particles should be relatively free. In the period in which Maxwell first worked, the mid-nineteenth century, this view was still controversial. Many people then believed that matter has a smooth, continuous basic structure, and it was not until late in that century that the atomistic concept became widely accepted. One of the reasons for rejecting the atomistic model was the apparent incompatibility of classical mechanics with the second law of thermodynamics and the obvious irreversibility of nature. In the 1860s, Maxwell made a major contribution when he took a first step to address this problem by introducing statistics to describe an atomistic gas. Specifically, he developed a function that describes the distribution of energies, and hence of velocities, of the moving particles comprising a gas, when they are in equilibrium at a fixed total energy as, for example, in a closed, insulated box. The velocities of individual particles could vary constantly, with each collision between particles or of particles with a wall. However the *distribution* of those velocities would remain essentially constant. There could be fluctuations in the distribution, of course, but these would be small and very transient.

Figure 20. James Clerk Maxwell.
(Courtesy of the Smithsonian Institution Libraries)

Maxwell's contribution was the "Maxwell distribution," or, as it is now called, the "Maxwell-Boltzmann distribution." Ludwig Boltzmann also derived that distribution and explored its implications extensively. This is still a mainstay in describing the most probable distribution of the energies of the molecules or atoms of any gas in equilibrium at moderate or high temperatures. Maxwell developed an expression for the probability that a molecule with mass m in a gas at temperature T has a speed between v and $v + dv$. This is an exponential; the probability is of the form

$$\text{Prob}(v) \sim e^{-K.E.(v)/kT} \text{ or, written alternatively, } exp[K.E.(v)/kT],$$

where the kinetic energy $K.E.(v)$ is $mv^2/2$ and e is the "natural number," approximately 2.718. (As we saw previously, the quantity in the exponent is called a "natural logarithm"; were we to write the expression as a power of 10, we would multiply that natural logarithm by the power to which 10 must be raised to get the value of e, approximately 0.435.) But the number of ways the speed v could be directed (in other words, the number of ways the velocity could be aimed) is proportional to the area of a sphere with radius v, so the probability of finding v must also be directly proportional to v^2. Without deriving the full expression, we can write the probability distribution so that the total probability, summed over all directions and magnitudes of the speed, is exactly 1:

$$Prob(v) = \sqrt{\left(\frac{m}{2\pi kT}\right)^3} (4\pi v)^2\, exp\left(-\frac{mv^2}{2kT}\right)$$

This is about the most complicated equation we shall encounter. This probability distribution rises from essentially zero for atoms or molecules with speeds close to zero, up to a peak, the most probable speed, which occurs at a speed $v_{most\ prob}$ of $\sqrt{2kT/m}$. Thus, the most probable speed increases with the square root of the temperature, and particles of small mass, such as hydrogen molecules or helium atoms, move much faster than heavier ones, such as nitrogen molecules. The nitrogen molecules have masses 14 times larger than those of hydrogen molecules, so the hydrogen molecules move roughly 3½ times faster than the nitrogen molecules. The probability distribution also tells us that there will be some, not many, particles moving at very high speeds, even at low temperatures. These will be rare, but they can be found nonetheless (Fig. 21).

Implicit in Maxwell's concept but not contained at all in the

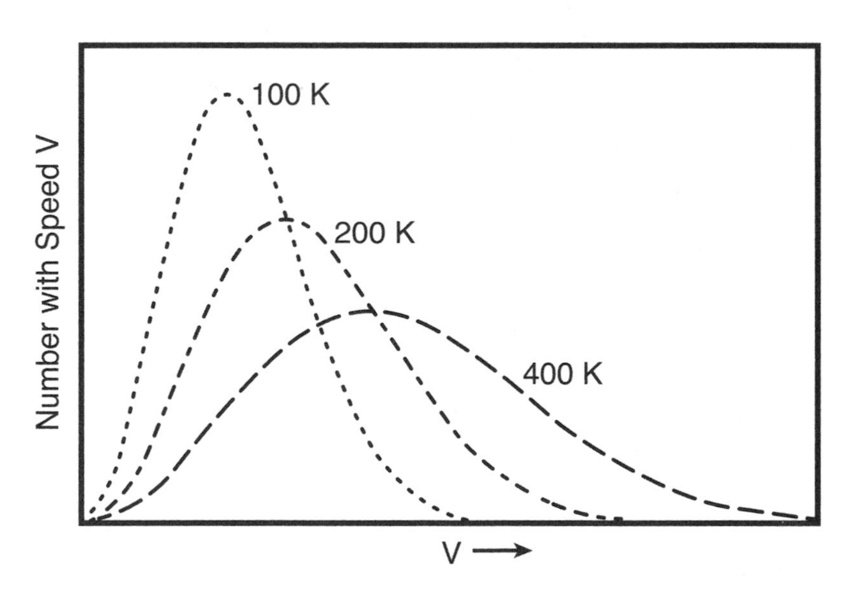

Figure 21. The shapes of the Maxwell-Boltzmann distribution of particle speeds at three temperatures. The peak moves to higher temperature and the area spreads to higher speeds, but the area under the curves is the same for all temperatures because the amount of matter is the same for all three of them. (Courtesy of Barbara Schoeberl, Animated Earth LLC)

expression for the probability distribution is the idea that there are fluctuations, both of the individual particle speeds and of the instantaneous average of their distribution. If the particles collide with the walls of their confining chamber, which they must, and with one another, which we now know they do (but Maxwell could only conjecture since he didn't know their size or shape), then the actual distribution of speeds changes constantly. However, the magnitude of the fluctuations of any system from its most probable distribution has a very interesting dependence on the number of particles in the system. If there are N individual particles in our system, then the probability of a deviation of the average speed or energy from its most probable value varies as $\sqrt{N}$ and hence the *fractional* probability varies as $\sqrt{N}/N$. Thus, if N is only 9, that fraction is 1/3, but if N is 1,000,000, the fraction is only 1/1000. And when we deal with macroscopic amounts of gases, we are working not with a million molecules but with roughly 10^{24} of them, so the chance of seeing a fluctuation away from the most probable distribution is about 1 in 1,000,000,000,000, hardly something we can hope to see.

We can be more precise about the deviations from the most probable by using the concept of *standard deviation from the mean*, especially for the most common and widely encountered form of distribution called a "Gaussian distribution," a symmetric, bell-shaped curve. A band within that curve that is one standard deviation wide from the most probable (and mean) value of that distribution either above or below that most probable value includes about 34% of the total population of the distribution, so about 68% of the total population lies within less than one standard deviation above or below the most probable value. Put another way, only about 1/3 of all the values of the distribution fall more than one standard deviation away from the mean, half of that third in the low tail and half in the high. And the deviation to which the above paragraph refers is specifically labeled as the *standard deviation.*

Maxwell arrived at the expression for the distribution of speeds by using plausible and intuitively acceptable ideas about the randomness of the particle velocities and their collisions. He did not find it by any set of logical steps or mathematical derivations. Carrying out that derivation mathematically was the first major contribution to this subject made by Ludwig Boltzmann. In 1872, Boltzmann started with the ideas that (a) the particles of a gas must be moving randomly because their container does not move, (b) their speeds, more strictly their velocities (which indicate both speed and direction), are randomly distributed at every instant, and (c) the particles collide with one another and, thereby, change speeds and directions with every collision, and the collisions happen very often. Using these completely plausible assumptions, he was able to derive, from a combination of physics, mathematics, and statistics, a distribution of velocities (and hence of speeds) for a gas in a state of equilibrium at a fixed temperature. And that distribution was exactly the one that Maxwell had produced intuitively! That distribution, the one given in the equation above and shown schematically in the accompanying diagram, is now known as the Maxwell-Boltzmann distribution, and is one of the fundamental links between the macroscopic approach of traditional thermodynamics and the microscopic approach based on the mechanics describing the behavior of matter at the atomic scale.

Boltzmann went further in linking thermodynamics to other aspects of science. He explained the basis of the connection, found experimentally by his mentor Josef Stefan, between the electromagnetic radiation emitted by a warm body and the temperature of that body; this relation, that the total rate of emission of electromagnetic energy is directly proportional to the fourth power of the body's temperature, is known as the Stefan-Boltzmann law. Moreover, Boltzmann showed that radiation exerted a pressure, just as a gas does. As we have seen, a warm object gives off radiation of very long wavelength, radio waves; a still warmer object loses energy by emitting microwaves; we know that very hot objects, such as flames and filaments in incandescent lamps, emit visible light. Still hotter objects, such as our sun, emit visible, ultraviolet, and even X-ray radiation.

One more very important contribution Boltzmann made to thermodynamics was a further application of statistics. He showed that if any collection of atoms with any arbitrary initial distribution of velocities is left in an isolated environment to evolve naturally, it will reach the Maxwell-Boltzmann equilibrium distribution. He also introduced a function he labeled "H" that measured how far that evolution to equilibrium has gone; specifically, he proved that function achieves its most *negative* value when the system comes to equilibrium. After Boltzmann first introduced this concept, it was called the H-function. The initial negative sign, however, was later reversed, so that the proof, applied to the redefined statistical H-function now called *entropy*, shows that function evolves to a *maximum*. And the proof was a microscopically based, statistically based demonstration that a system's *entropy* takes on a maximum value when it is in equilibrium. The H-function was precisely the micro-based equivalent and counterpart of the macro-based entropy, as developed by Clausius, but with the opposite sign.

The next major contribution to the linking of thermodynamics and mechanics came from J. Willard Gibbs. Largely unrecognized in the United States until European scientists read his works and realized their import, Gibbs established the field of statistical thermodynamics on a new, deeper foundation. His approach to applying statistics was based on the then novel concept of *ensembles*, hypothetical large collections of replicas of the system one is trying to describe. For example, if one wants to understand the behavior of a system in equilibrium at a constant temperature, presumably by its contact with a thermostat, then one imagines a very large collection of such systems, all with the same macroscopic constraints of constant temperature and presumably volume and pressure, but each system with its own microscopic assignments of molecular or atomic properties that would be consistent with those macroscopic constraints. Thus the individual systems are all different at the microscopic level, but all identical at the level of the macroscopic, thermodynamic variables. Then the properties one infers for the real system are the *averages of those properties over all the systems of the ensemble.* Constant-temperature ensembles, also called "canonical ensembles" or "Gibbsian ensembles," are not the only useful ones; others, such as constant-energy ensembles, called "microcanonical ensembles," are used to describe the behavior of isolated systems, for example. The energies of the systems in the constant-temperature ensemble are not all the same; they are distributed in a manner analogous to the Maxwell-Boltzmann distribution of velocities in a single constant-temperature system. Other kinds of ensembles also come into play when one introduces other constraints, such as constant volume or constant pressure.

A very specific product of Gibbs's work was the famous phase rule, often called the Gibbs phase rule, which we discussed in Chapter 2. We shall return to this later, when we investigate the boundaries of applicability of thermodynamics. It was Gibbs who introduced the term "statistical mechanics" and the concepts of chemical potential (free energy per unit of mass) and of phase space, the six-dimensional "space" of the three spatial position coordinates and the three momentum coordinates.

Thermodynamics and the New Picture of Matter: Quantum Mechanics

A revolution in our concept of the nature of matter—at the microscopic level—began in 1900 when Max Planck, in explaining the distribution of spontaneous radiation from any object at a fixed temperature, proposed that energy as radiation should come in discrete units, "quanta," rather than in a continuous stream. The previous ideas about radiant energy were clearly incorrect, because they predicted that any object would radiate ever increasing amounts of energy of ever smaller wavelengths, implying that the energy being radiated would be infinite, and predominantly of infinitesimal wavelength. This was obviously impossible and wrong, and came to be known as "the ultraviolet catastrophe," but it was not clear, until Planck, how to correct it. Requiring that radiation comes in discrete quanta, and that the energy in each quantum be directly proportional to the frequency of the radiation (and hence inversely proportional to its wavelength) resolved that discrepancy, so that the amount of radiation from any object diminishes with increasing frequency, beyond some frequency where the radiation is a maximum. Obviously, since an object glowing dull red is not as hot as one glowing bright yellow, the maximum frequency of the emitted radiation, as well as its total amount, increases with the object's temperature.

A major advance came soon after, when, in 1905, Einstein showed that radiation, with the form of discrete quanta, explained the photoelectric effect, by which electrons are ejected from a substance when it is struck by light of sufficiently short wavelength. It was for this discovery that he received the Nobel Prize—and he later said that, of the three major discoveries he published that year (special relativity, the explanation of Brownian motion, and the origin of the photoelectric effect), explaining the photoelectric effect was the one that truly deserved the prize. It was in that analysis that Einstein showed that excited atoms or molecules could be induced to emit radiation and reduce their energy if they were subjected to the stimulus of incident radiation of precisely the same frequency as that which they would emit, a process, now called "stimulated emission," which is the basis for lasers.

Then, the next step came when Niels Bohr devised a model to explain the radiation in the form of spectral lines coming from hydrogen atoms. He proposed that the one negatively charged electron, bound to a much heavier, positively charged proton, could only exist in certain specific states, each with a very specific energy, and that a light quantum corresponding to a specific spectral wavelength is emitted when an electron falls from a specific state of higher energy to one of lower energy. Likewise, when a quantum of that specific energy is absorbed by an electron in its lower-energy state, that electron is lifted to the state of higher energy. The Bohr model has a lowest possible energy for the bound electron, and an infinite set of higher-energy bound levels that converge at just the energy required to set the electron free of the proton. Above that level, the electron can absorb or emit energy of any wavelength; below that limit, only quanta of energies corresponding to the difference between two discrete levels can be absorbed or emitted. Extending the Bohr model to atoms with more than one electron proved to be difficult, until the discovery of the much more elaborate, sophisticated subject of "quantum mechanics," by the German physicists Werner Heisenberg and Irwin Schrödinger in 1924, and its further elaboration by many others. Now we use quantum mechanics to describe virtually all phenomena at the atomic level; the mechanics of Newton remains valid and useful at the macroscopic level of real billiard balls and all our everyday experience, but it fails badly at the atomic level, where quantum mechanics is the valid description.

What does quantum mechanics have to do with thermodynamics? After all, quantum mechanics is relevant at the atomic level, while thermodynamics is appropriate for macroscopic systems. One context in which quantum mechanics becomes especially relevant is that of systems at low temperatures. There, one of the most important manifestations comes from the striking property that quantum mechanics reveals regarding the fundamental nature of simple particles. Specifically, quantum mechanics shows us that there are two kinds of particles, called bosons (after the Indian physicist S. N. Bose) and fermions (after the Italian-American physicist Enrico Fermi). Electrons are fermions; helium atoms with two neutrons and two protons in their nuclei are bosons. The distinction relevant here between these two is that there is no limit to the number of identical bosons that can be in any chosen state, whereas only one fermion can occupy a specific state. When one fermion is in a state, it is closed to all other identical fermions. This means that if we cool a system of bosons, which has a lowest state and states of higher energy, withdrawing energy from that system by lowering its temperature more and more, the individual bosons can all drop down and, as the system loses energy, more and more of the bosons can fall into the state of lowest energy. The fermions, in contrast, must stack up, since a state, once occupied by one fermion, is closed to any others. This means that there is a qualitative difference between the nature and behavior of a system of bosons and a system of fermions, a difference that reveals itself if the energies of these systems are so low that most or all of the occupied energy states are their very low ones.

Because of their discrete, quantized levels, atoms, molecules, and electrons can exhibit properties very different at extremely low temperatures from those of matter at room temperature. Bosons assembled in a trap, for example, can be cooled to temperatures at which most of the particles are actually in their lowest energy state; a system in such a condition is said to have undergone Bose or Bose-Einstein condensation. Fermions, likewise, can be cooled to a condition in which many or most of the particles are in the lowest allowed states in the stack of energy levels.

There is an interesting implication from this behavior regarding the third law of thermodynamics. Normally, we can distinguish systems of constant temperature—called "isothermal" systems—from systems of constant energy, called "isolated" or "isoergic" systems. We also distinguish the Gibbsian ensembles of systems at a common constant temperature—"canonical ensembles"—from ensembles of systems all at a constant common energy, or "microcanonical ensembles." Recall that the third law of thermodynamics says that there is an absolute lower bound to temperature, which we call 0 kelvin, or 0 K. Moreover, that law states that it is impossible to bring a system to a temperature of 0 K. However, if we could bring an assembly of atoms to its lowest possible quantum state, we would indeed bring it to a state corresponding precisely to 0 K. Whether that can be done in a way that the particles are trapped in a "box," a confining region of space *and are all in the lowest possible energy states allowed by the confines of the box*, is an unsolved problem, but if it can be done, and that may be possible with methods now available, then that system would be at 0 K, in apparent violation of the third law. However, even if it is possible to produce such an ultracold system, we recognize that this could be done only for a system composed of a relatively small number of particles, certainly not of a system of very tiny but macroscopic size, say of 10^{15} atoms. In other words, we may expect to see a violation of one of the laws of thermodynamics *if we look at sufficiently small systems.* Again, as with entropy always increasing, we could see transient fluctuations violating the second law if we look at systems so small that those fluctuations are frequent and large enough to be detected. This simply emphasizes that thermodynamics is truly a science of macroscopic systems, one that depends in several ways on the properties of large numbers of component particles.

The concept of a real system attaining a temperature of 0 K introduces another sort of paradox that distinguishes small systems from those large enough to be described accurately by thermodynamics. Traditionally, the two most widely used concepts of Gibbsian ensembles are the canonical or constant-temperature ensemble, and the microcanonical or constant-energy ensemble. Constant-temperature systems show very different behavior in many ways from constant-energy systems, so we naturally recognize that the canonical and microcanonical ensembles we use to describe them are, in general, quite different. However, if we think of an ensemble of systems that have all been brought to their lowest quantum state, those systems have a constant energy and hence are described by a microcanonical ensemble—yet they are all at a common temperature of 0 K, and hence are described by a canonical ensemble. Thus we come to the apparent paradox of a special situation in which the canonical and microcanonical ensembles are equivalent. Once again, we arrive at a situation that doesn't fit the traditional rules of classical thermodynamics, but only for systems so small that we could actually bring them to their lowest quantum state.

We thus recognize that, valid as it is for macroscopic systems, thermodynamics is an inappropriate tool to describe very small systems, systems composed of only a few atoms or molecules. Modern technologies give us tools to study such small systems, something quite inconceivable in the nineteenth or even much of the twentieth century. This new capability opens a new class of questions, questions that are just now being explored. Specifically, we can now ask, for any given phenomenon whose thermodynamic description is valid for macroscopic systems but fails for very small systems, "What is the approximate size of the largest system for which we could observe failure of or deviations from the thermodynamics-based predicted behavior?" This brings us to our next chapter.

SIX

How Can We Go Beyond the Traditional Scope of Thermodynamics?

We have already seen that there are some situations in which general properties of macroscopic systems, well described by thermodynamics, lose their applicability when we try to apply those ideas to small systems. One example we have seen is the possible violation of the third law, for collections of atoms or molecules so few in number that those particles might all be brought to their lowest quantum state.

Another is the Gibbs phase rule; recall that the number of degrees of freedom f depends on the number of components c or independent substances, and on the number of phases p in equilibrium together: $f = c - p + 2$, so that water and ice, two phases of a single substance, can have only one degree of freedom. Hence at a pressure of 1 atmosphere, there is only one temperature, 0° C, at which these two phases can be in equilibrium.

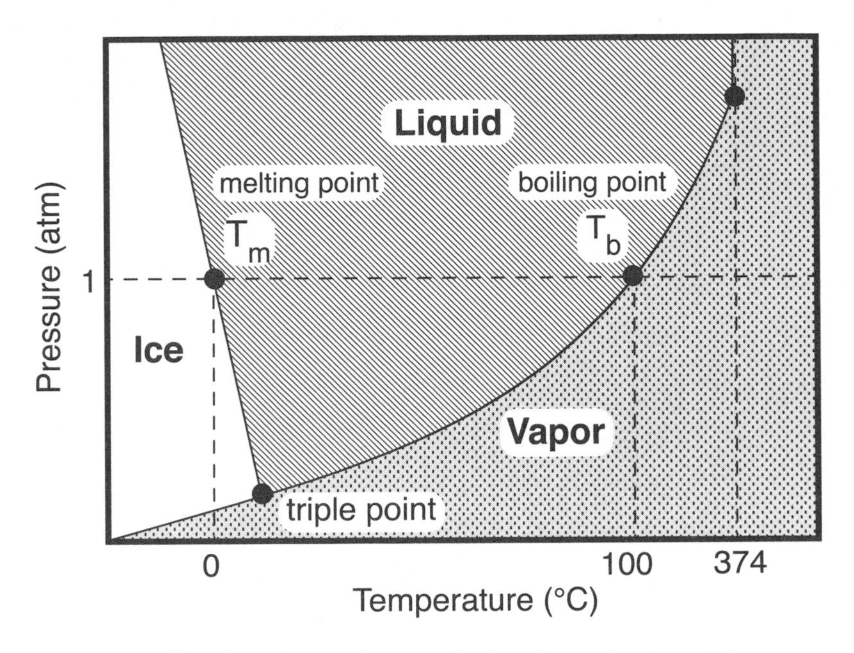

Figure 22. The phase diagram for water. In the white region, ice, solid water, is stable. In the striped region, liquid water is stable. In the dotted region, the stable form is water vapor. Ice and liquid water can coexist in equilibrium under conditions along the boundary separating the white and striped regions. Liquid water and vapor can coexist in equilibrium along the curve separating the striped and dotted regions. The three forms or phases can coexist together under only one temperature-pressure pair, labeled as the "triple point." (Courtesy of Barbara Schoeberl, Animated Earth LLC)

The "phase diagram" for water illustrates the regions of temperature and pressure in which solid, liquid, and gaseous water can exist as stable forms (Fig. 22). Phases can coexist at the boundaries between regions. Thus, if we change the pressure, the temperature at which water and ice can coexist changes. This rule describes the behavior of coexisting phases of macroscopic systems, and seems to be a universally true condition—*for macro systems.*

If we look at the behavior of small clusters, however, say 10, 20, or 50 atoms, we find, both in experiments and computer simulations, that solid and liquid forms of these tiny beasts may coexist in observable amounts within observable ranges of temperature and pressure! Their coexistence is not restricted to a single temperature for any given pressure. This reveals that small clusters of atoms and molecules simply do not obey the Gibbs phase rule. But if we look carefully at the fundamental thermodynamic foundations of equilibrium, this apparent anomaly becomes quite easy to understand, and is not actually a violation of fundamental thermodynamics.

If two forms of a substance, A and B, are in equilibrium, as we have seen, the ratio of their amounts [A] and [B] is the temperature-dependent equilibrium constant $K_{eq} = [A]/[B] = \exp(\Delta G/kT) = \exp(N\Delta\mu/kT)$. We want to use that final form because we'll examine what this expression tells us both when N has a value corresponding to a macroscopic system and when N corresponds to just a few atoms or molecules. Suppose that the free energy difference between two phases, per atom, $\Delta\mu/kT$, is very tiny, say 10^{-10}, so that kT, effectively the mean thermal energy per particle, is 10^{10} times larger than that difference in free energies per atom or per molecule. And suppose we are dealing with a small but macroscopic system, only 10^{20} particles, less than a thousandth of a mole, roughly a grain of sand. In this case, the equilibrium constant, the ratio of the two forms of the substance, is either $\exp(10^{10})$ or $\exp(10^{-10})$, either an enormously large or infinitesimally small number, depending on which of the two phases is favored by having the lower free energy. This tells us that even just this very tiny bit away from the conditions of exact equality of the free energies and chemical potentials of the two forms, the minority species would be present in such infinitesimal amounts that it would be totally undetectable.

It is precisely this character that makes phase changes of ordinary substances sharp, effectively like sudden, abrupt changes with temperature at any fixed pressure. In principle, phase transitions are smooth and continuous, but for all practical purposes we see them as sharp changes, so that we can treat them as if they were discontinuous—for macroscopic systems. This is why we say water has a sharp, specific melting temperature and a sharp, specific boiling temperature at any chosen pressure.

Now let us apply the same kind of analysis to small clusters. Suppose our system consists of only 10 atoms. That is, suppose N = 10, not 10^{20}. In that case, K_{eq} can easily be close to 1 over a range of temperatures and pressures. For example, if $\Delta\mu/k\mathrm{T}$ is ± 0.1, meaning the free energy per atom of the favored form is 10 times lower than that of the unfavored, then $K_{eq} = \exp(\pm 0.1 \times 10) = \exp(\pm 1)$ or ca. 2.3 or 1/2.3. This means that it would be easy to observe the less-favored form in equilibrium with the more favored when the free energies per atom of the two forms differ by a factor of 10—which is a clear violation of the Gibbs phase rule. Hence this apparent anomaly is entirely consistent with *fundamental* thermodynamics. We see that the Gibbs phase rule is a rule about macroscopic matter, not about atomic-size systems. We can go further, if we know about how large the fraction of a minority species would be detectable experimentally. We can find how large the size a cluster of atoms would be that would just allow us to detect, say, a liquid at a temperature below its formal freezing point, for example 0° C for water, and also what the lower limit of temperature would be of the band of coexisting liquid and solid. We learn, for example, that our experiments could reveal solid and liquid clusters of argon atoms coexisting over a few degrees Celsius, if the clusters consist of 75 or fewer atoms. But if the clusters were as large as 100 atoms, their coexistence range of temperature would be too narrow for us to detect anything but a sharp transition with the tools we have today.

Thermodynamics and Equilibrium

Thermodynamics, as we have examined it, concerns itself with macroscopic systems, with the equilibrium states of those systems, with idealized changes among those equilibrium states, and, by treating those idealized changes as natural limits, with the natural bounds on performance of real systems that execute real changes among states. The ideal, reversible heat engine, whether it follows a Carnot cycle or some other cyclic path, must have an efficiency, the amount of work produced per amount of heat taken from a hot source, partly using that heat energy to produce work and part to deposit in a cold sink. That is precisely what Carnot taught us: the maximum possible efficiency is $1 - (T_L/T_H)$ where T_L and T_H are the temperatures of the cold sink and the hot source, respectively. The infinitesimal steps that the ideal engine takes to go through its cycle are really intended to be steps from one equilibrium condition to the next. The very nature of the idealized *reversible* cycle is one of a succession of hypothetical equilibrium states; that is the essence of reversibility. We can never tell which way the system came to its present state or which way it will go next, just by looking at the system and not at its history. Real engines are of course never in equilibrium when they are operating; we describe them in a useful, approximate way by attributing a pressure, a volume, and a temperature to our engine at every instant along its path, but strictly, it is constantly changing its conditions and is never in equilibrium while it operates. We make a fairly accurate, useful approximation when we attribute a single pressure, volume, and temperature to the real engine at every instant, but strictly, we know that it is an approximation, a fiction. But it's accurate enough to be very useful.

That traditional thermodynamics is about states in equilibrium and the changes associated with passages between these states has long been accepted. In effect, there were tacit limits to the domain of applicability and validity of thermodynamics; the systems for which traditional thermodynamics can be used should always be close to equilibrium. Well into the twentieth century, however, those limits were challenged, and the subject of thermodynamics began to be extended to systems that are not in equilibrium. We shall not go at all extensively into these extensions here, but merely give a brief overview of what they do and how they can be used.

The first step, a very major one, was the introduction of flows as variables. Clearly, describing any system not in equilibrium must require additional variables for its characterization. Lars Onsager, a professor of chemistry at Yale, formulated a way to describe systems near but not in equilibrium by finding relations among the various flows, the rates at which mass of each specific kind, and heat, and whatever other variables undergo changes in the system. A system out of equilibrium is more complex than one in equilibrium, so we need to use more information to describe it and its complexity. The rates of flow provide just that additional information in the formulation developed by Onsager. These additional variables augment the traditional variables of thermodynamics. Of course systems that are not in equilibrium typically have different values of temperature, for example, at different places within them. They often have different relative amounts of different constituent substances as well. But in the context in which Onsager's approach applies, we can associate values of the traditional thermodynamic variables with each point, each location, within the system. We call that condition "local thermal equilibrium," often abbreviated LTE.

With that as a stimulus, other investigators opened new ways to extend thermodynamics to treat systems operating at real, nonzero rates. For example, one productive direction has been the extension of the concept of thermodynamic potentials to processes operating in real time. As we discussed earlier, traditional thermodynamic potentials are the functions whose changes give the absolute natural limits on performance, typically the minimum energy required to carry out a specific process with an ideal, reversible machine, one that operates infinitely slowly. Each set of constraints such as constancy of pressure or volume or temperature, or combination of any two of these, calls for its own thermodynamic potential. The extension to real-time processes requires more information, of course. Specifically, the analogues of traditional thermodynamic potentials for real-time processes require knowledge of whatever totally unavoidable loss processes must be accepted and hence included in the way the process is specified. For example, if a process involving a moving piston is to be carried out in a specified time, it is likely that the friction of the piston against its container's walls would be an unavoidable loss of energy, something that would have to be included in the appropriate thermodynamic potential for the process. These extended limits on performance are naturally called "finite-time potentials."

Another aspect of the extension of thermodynamics to finite-time processes is the recognition that, other than efficiency—work done, per total energy put into the process—there is at least one other quantity that one may wish to optimize, namely the *power* of the process, the amount of useful energy delivered, per unit of time. This was first investigated by two Canadians, F. L. Curzon and B. Ahlborn; they compared the optimization of a heat engine for maximum power with that for maximum efficiency. In contrast to the Carnot expression for maximum efficiency, $1 - T_L/T_H$, its efficiency when a machine delivers maximum power is given by $1 - \sqrt{T_H/T_L}$, with the square root of the temperature ratio replacing the simple ratio of temperatures. Of course the pathway that yields maximum power is very different from the maximum-efficiency path, the infinitely slow reversible path, since power requires that the energy be delivered at a real, nonzero rate. Other kinds of thermodynamic optimizations can be useful for real, finite-time processes as well, such as minimization of entropy production. That choice uses a measure of "best way" different from both the maximum efficiency and the maximum power criteria. Each of these leads to a different "best" pathway to achieve the desired performance. But all fall well within the spirit of thermodynamics, its reliance on macroscopic properties and variables, and its goal of optimizing performance, by whatever criterion.

There is still another important step one must carry out in order to design and construct real processes aimed toward achieving or approaching the "best possible" performance, whatever criterion of "best" one chooses. That is finding the pathway, typically of a cyclic or continuous flow process, that comes as close as possible to meeting the chosen criterion of "best" performance. This is achieved by a mathematical approach known as "optimal control theory." Finding extremal pathways, such as longest or shortest, began in the 1870s with Johann Bernoulli and other mathematicians, and continued as an active topic, perhaps culminating in the work of the blind mathematician Lev Pontryagin and other Russians in the 1950s and '60s. The formal mathematical methods they introduced can now be supplemented or even replaced by computational simulations, enabling one to design process pathways that bring the process as close as the available technology allows to whichever optimal path one wishes.

Sometimes thermodynamics opens realizations that something can be treated as a controllable variable that was not previously recognized as such. Distillation, for example, one of the most widely used industrial heat-driven processes, common in petroleum refining and whisky making, is traditionally employed the same way as it is in undergraduate laboratory courses, with the heat put in entirely at the bottom of the distillation column, where the material to be distilled lies, while the column above that container is cooled by a steady flow, typically of cold water. The most volatile component of the original mixture rises to the top of the column more readily than any less volatile component, but only by going through many steps of condensation and re-evaporation all along the height of the column. It turns out that much waste heat goes into that re-evaporation. By treating the temperature profile of the column above the source at the bottom as a control variable, one can minimize the heat input required to achieve the desired separation. This is an example of the application of optimal control to achieve maximum efficiency in distillation.

One can carry out analogous optimizations to maximize power or to minimize entropy production, for example, for other processes. A key step and often an intellectual challenge is identifying and choosing the right *control variable*, the variable whose values we can choose; once selected, we can go through a mathematical analysis to find the pathway for those values that yields the best performance, based on the criterion we have chosen to optimize. The temperature profile of the distillation column is just one example, specific to that situation; each process we may wish to optimize will have its own control variable or variables. Of course it is important to choose as one's control variable something that can truly be controlled in practice!

In its present state, the extension of thermodynamics to nonequilibrium systems is a useful tool for describing systems for which traditional variables—temperature, volume or density, pressure—can be applied, supplemented by others such as flow rates. However, if a system is so far from equilibrium that those traditional variables are changing at a rate so fast that they are not useful, as in combustion, for example, the methods of thermodynamics, even as they have been developed and extended, are not applicable. For those phenomena, we must find other approaches, some of which are well understood—for example, much of traditional kinetics of chemical reactions—and others, including several aspects of reaction kinetics, that literally lie on the frontier of current research.

SEVEN

What Can Thermodynamics Teach Us About Science More Generally?

Thermodynamics as a unified science evolved initially from the stimulus of a very practical effort to make heat engines more efficient, to get as much work from heat as we can. The challenge to mine operators was to spend as little money as possible for the fuel required to run the steam engines that pumped water out of their mines. Watt's engine and Carnot's analyses came from that search. Only after their contributions and those of their contemporaries and immediate successors benefited the mining community did the basic science emerge. The solution to this engineering problem provided the understanding of the nature of heat and the recognition of the interconvertibility of light, heat, mechanical work, and then electricity. With these understandings, of course, came the explicit idea of energy. We have seen how it was this interconvertibility that eventually led to the unifying concepts of energy and entropy, perhaps painfully, as the central foundations of a powerful science. Here is a paragon of the basic, general, and unifying arising from the stimulus of specific applied problems.

This is an interesting contrast to a widely recognized and accepted paradigm of the twentieth and twenty-first centuries, that the applied follows as a consequence of the basic. That model has many examples to demonstrate its validity. What we learn from the evolution of thermodynamics, however, is that concepts and their consequences can flow in either direction, the basic out of problems of application, or all sorts of applications from recognition of potentialities exposed by new basic knowledge. The laser is a consequence of Einstein's analysis of radiation, and of the necessity that light of a precise frequency can stimulate an appropriate energized particle to emit still more light of that frequency—a case of basic science leading us to what has become an extremely important application. The determination of the helical, sequential structure of DNA is another example of a discovery of a "pure" basic character eventually yielding new, practical methods to carry out genetic control. So here is a way that thermodynamics can broaden our perspective on how scientific advances come about; the basic can lead to the applied, or, with thermodynamics, the applied can lead to the basic.

Variables and Ways of Looking at Nature

Classical and quantum mechanics approach the description of nature in terms of the behavior of individual elements, whether atoms or tennis balls or planets. The traditional variables we use for these in both approaches are the ones describing the position and motion of those elements, how they change and how they influence those properties of each other. This is what we have called the "microscopic approach," for which we have adopted variables of position, velocity, acceleration, momentum, force, and kinetic and potential energy. But that approach is of course not restricted to truly microscopic things. It is equally valid for describing individual objects such as billiard balls and baseballs. It is applicable to atoms and baseballs because we characterize them using properties such as position and velocity. Thermodynamics and statistical mechanics take a very different pathway to describe nature. They use properties we associate with everyday objects and aspects of our environment as the variables telling us about the world we inhabit. Temperature and pressure are properties of complex systems, typically composed of many, many of the elements that mechanics treats individually. Temperature and pressure have no meaning for individual elements; momentum and velocity have no meaning for a complex system, particularly when it is in equilibrium, unless we treat the entire system as an element, such as a tennis ball.

It is amusing, in a way, to recognize that the tennis ball can be considered as a single element whose complex molecular structure we totally neglect, for which mechanics provides the appropriate description, or as a complex of molecules with a temperature and an internal pressure, whose velocity and momentum we neglect, when we use the appropriate thermodynamic, macroscopic description. When treated as a single object, we can ascribe to it a mass, an instantaneous position of the center of that mass, a velocity and an acceleration of the center of that mass, and still other characteristics such as the compressibility of the tennis ball and the instantaneous distortion from spherical shape when the ball strikes a surface. On the other hand, we could, in principle, describe the tennis ball as composed of a vast number of molecules bound together to form a stable but somewhat elastic object. But in practice, we cannot play this two-faced game of macro and micro descriptions with individual molecules or with the steam driving an engine; we must use the microscopic approach for the former and the macroscopic, thermodynamic approach for the latter. Of course, if we go exploring into the nature of the particles of which atoms are made, we come to a new level of description. Protons and neutrons themselves consist of particles, called quarks and gluons. Is there another level, still deeper, of which these are made? We have no evidence now of such, but we cannot predict what people might find in the future. The history of science tells us that we continue to invent new tools to probe ever deeper into the structure of matter; where this journey will take us in the future, we cannot predict.

Statistics—as statistical mechanics and statistical thermodynamics—provides a bridge to link these two very different modes. By recognizing that we cannot imagine describing the instantaneous velocities and positions of all the air molecules in a room, but that they nevertheless do obey the laws of mechanics, quantum or classical, and then introducing the extremely critical concept of *probability*, we can ask, "What are the most probable conditions, in terms of microscopic variables, of those many air molecules, if we specify the macroscopic variables that describe what we observe?" And then we must ask, "What is the probability that we could observe a measurable deviation from those most probable conditions? How probable is the most probable, compared with other possible conditions?" As we saw in Chapter 4, the behavior of systems of large numbers assures that we have virtually no possibility of observing a significant deviation from the most probable condition for a typical macroscopic system; we will never see all the air molecules moving, even very, very briefly, to one side of the room, leaving none on the other side. But if we consider, say, only ten air molecules in a box of one cubic centimeter, it is quite likely that we could sometimes find all ten on one side of that box. The more elementary objects that the system comprises, the more unlikely and difficult it is to observe deviations from the most probable macrostate of that system.

This way of using statistics and probability tells us that our macroscopic description is indeed valid, that the macro variables we use are indeed the appropriate ones for the macro world in which we live. Because deviations from the most probable conditions have probabilities that depend on the square root of the number of elements that compose the macro system, and because square roots of very large numbers are very much smaller than those large numbers, the deviations become too improbable for us to expect to observe them—ever.

There is another way to use that information, however. There are some kinds of behavior of macroscopic systems that we know and understand and accept as universally correct, but which become inaccurate and inappropriate for small systems, for which fluctuations from the most probable can be readily observed. A striking example that we have seen is the case of the Gibbs phase rule, which says that while we can vary the temperature and pressure of a liquid, say a glass of water, as we wish (within limits, of course), if we insist on having ice in the water, with the two forms in equilibrium, there is only one temperature, 0° C, at one atmosphere pressure, at which we can maintain those two phases together. If we change the pressure, then the temperature has to change in order for the two phases to stay in equilibrium. If we want an even stricter condition, in which the ice, the liquid water, and water vapor are all three in equilibrium, there is only one temperature and pressure at which that can happen. This is the content of that phase rule, $f = c - p + 2$. The condition for different phases to be in equilibrium is simple to state: they must have exactly equal free energies, per particle or per mole, or, in other words, exactly equal chemical potentials.

If we look at the properties of very small systems, for example a cluster of only 20 or 50 atoms, we see something very different from what the phase rule tells us. We have seen that solid and liquid forms of these small but complex particles can coexist within a range of temperatures and pressures. Moreover, more than two phases of these clusters can coexist within a finite range of conditions. At the point at which the free energies of the solid and liquid forms are equal, the probabilities of finding solid or liquid are exactly equal. If the free energies of the two are different, however, the phase with the lower free energy is the more probable, but the other, the "less favored phase," is present as a minority component of the total system—in *observable amounts.*

How can this be? How can small systems violate the presumably universal Gibbs phase rule? The answer is straightforward and lies in the properties of systems of large numbers. The ratio of the amounts of two forms of a substance that can coexist in equilibrium is determined by the exponential of the free energy difference between the two forms. The exponent, that free energy difference, must of course be in dimensionless units; we cannot take *e* or 10 to a power of a dimensioned quantity. The free energy must be expressed in units of the energy associated with the temperature of the system, which, as we have seen, we can write as *k*T if the free energy difference is expressed in a per-atom basis or as *R*T if in a per-mole basis. If, for example, substance A and substance B could interconvert, and at a specific temperature and pressure, their free energy difference per molecule, divided by *k*T, is equal to the natural logarithm of 2, then the equilibrium ratio of the amount of A to the amount of B would be $e^{\log 2}$, just 2:1. However, if our system is a small but definitely macroscopic one, say of 10^{20} particles, roughly the size of a grain of sand, and the free energies of the two phases differ by a very tiny amount, say 10^{-10} in units of *k*T, the ratio of the two phases is $\exp\{\pm 10^{10}\}$, either an enormous number or an extremely tiny number. This tells us that although, in principle, the two phases could coexist in unequal amounts under conditions where they have different free energies, those coexistence ranges would be far too narrow for us ever to observe them. What is formally a continuous change from solid to liquid, for example, occurs over a range of temperatures (at constant pressure) so narrow that it appears to be a sudden, discontinuous change by any means that we have to observe such a change. Only at the point of exact equality can we see the two phases in equilibrium—in practice.

If, however, we look at small systems, the free energy differences between the solid and liquid phases of a 20- or 50-atom cluster can be small enough that the exponential of that difference can be near but different from 1, meaning that the unfavored phase, the one with the higher free energy, can easily be present as a minority species *in observable amounts.* In short, the phase rule is absolutely valid for macroscopic systems, because of the properties of large numbers, but it loses its validity for systems of small numbers of component particles because the large-number dominance doesn't hold for those systems. In fact, it is even possible to estimate the largest size of systems for which this violation of the phase rule would actually be observable. Clusters of a hundred or so metal atoms can exhibit coexisting solid and liquid phases over an observable range of temperatures, with the relative amounts of the two phases governed by the difference in the free energies of those phases. However, although we could observe coexisting solid and liquid clusters of 75 argon atoms within a range of temperatures, such a range of coexistence would be unobservably narrow for a cluster of 100 argon atoms.

More broadly, this illustration is meant to show that there can be macroscopic phenomena that are fully and accurately described by our macroscopic scientific methods, which lose their validity for small systems, simply because that validity is a consequence of the behavior of large numbers. We can generalize the conclusion of the previous paragraph by pointing out that, for any property described accurately by our macroscopic approach but which fails for small enough systems, it is possible at least in principle to estimate the size of systems that fall at the boundary, the lower size limit for validity of the macro description, below which we must use some more microscopic approach. In the case of the failure of the phase rule, we can, in fact, continue to use some aspects of the macroscopic approach because we can continue to associate a free energy with the atomic clusters of each phase, and to use the exponential of that difference to tell us the relative amounts of each that we can expect to observe. However, we may also want to ask, and this is at the frontier of the subject, what fluctuations from the most probable distribution we can expect to observe. In a sense, we can define a boundary size for a property well described at the macro level that, for small systems, does not conform to that description; we can ask, and sometimes estimate, the largest size system for which we would be able to observe behavior deviating from the macro description. We can ask, for example, what would be the largest number of molecules in a box of given size, for which we could actually observe, briefly of course, all the molecules on one side of the box. But finding such "boundary sizes" for specific phenomena is just now at the frontier of science.

This tells us something more general about science. We find concepts, sometimes from everyday experience, such as temperature, pressure, and heat, that we can learn how to relate to one another. But we also seek to give some kind of deeper meaning to those concepts so we may propose ideas, notions we might not (yet) be able to observe or confirm, such as the proposal that matter is composed of atoms, or that protons are composed of quarks and gluons, or that heat consists of a fluid we call caloric. There was a time, well into the nineteenth century, when the existence of atoms was only a conjecture, and a very controversial one—but when Maxwell and Boltzmann showed how one could derive many observable properties from the assumptions of the existence, motion, and collisions of atoms, people became much more willing to believe in their existence. And when Einstein explained Brownian motion, the random, fluctuating movement of small particles visible under a microscope, in terms of the collisions of those particles with molecules, the acceptance of atoms and molecules became virtually universal.

This is an example of how we developed tools to test proposed explanations. In this case, eventually we were able to go so far as to verify the atomic hypothesis by actually seeing the proposed objects, and, in fact, not only to see but even to manipulate atoms. With tools to work at the scale of ten billionths of a centimeter, we now are able to move and place individual atoms. But for many years, we lived with the unverified belief (unverified at least in terms of actually observing them) that atoms are the basic constituents of the matter we see all the time. Of course the observation in 1827 by Robert Brown that tiny particles, pollen grains, observable under a microscope, jiggled and wiggled in random ways we call "Brownian motion" was especially strong evidence in support of the idea of atoms; something had to be hitting those particles we observed to make them move randomly, in small steps. It was Einstein who, in 1905, showed that random collisions of molecules with those observable pollen grain particles accounts very well for those jiggles. Now, on a finer scale of size, we see strong evidence that protons are not elementary particles but are made from smaller things, the things we call "quarks" that are held together by things we call "gluons." But we have not *seen* quarks or gluons, not yet. In contrast, we believe that electrons are not made of smaller things, but are actually elementary particles.

Hence, just as we have seen how thermodynamics came to be linked via statistical mechanics to atomic theory, other sciences evolve as we propose, test, and develop new tools to probe new concepts. Some of those tools are experimental, the results of new tools or new ways to use what we already have; others are new concepts, new ways to ask questions or to envision and hypothesize how things might be happening, concepts that lead to elaborate theories that can only achieve validation, ultimately, by experiments or observations. The gravitational force exerted by the mysterious dark matter is an example of a major puzzling challenge arising from observations. We see the consequences of that gravitational force, but we have not yet seen what is responsible for producing that force, so we don't yet know what its source is. Observing results consistent with or predictable from a theory only says that the theory *may* be correct, of course, and the more experiments and observations consistent with that theory, the more we are ready to accept and believe it. On the other hand, one experimental result or observation whose validity we accept that contradicts the predictions of the theory is enough to disprove the correctness of the theory. In science, it is often far easier to show that an idea is wrong than that it is correct. A single (but confirmable) example that contradicts a theory or an idea is enough to disprove it.

Nonetheless we can often go far with concepts and theories that turn out not to be universally valid, but that have clear, justifiable validity for some situations. This is exactly the case with thermodynamics; it is a valid science for systems composed of very many elementary components, but some of its "rules" and predictions lose their validity when we deal with systems of only a very few components. (Of course some aspects we think of as coming from thermodynamics do remain valid for all systems, such as the conservation of energy.) The same kind of limitation is true of the mechanics of Newton; it is quite valid and useful for describing baseballs or planets, but loses its validity at the very small scale of atoms, where we must use quantum mechanics. Nonetheless, in both of these examples, we now understand how the large-scale science, whether thermodynamics or Newtonian mechanics, evolves in a self-consistent way, from the small-scale science, how statistical mechanics explains the way thermodynamics becomes valid for many-body systems and how quantum phenomena become unobservable as we move from the atomic scale to the scale of human beings.

Sometimes we can demonstrate that a proposed concept is wrong; that is what happened with the caloric theory of heat, and that process of negation was the main pathway for the evolution of science. Negation has always been one of science's strongest tools; it is usually far easier to show that a wrong idea is wrong than that a correct idea is indeed right. This is obviously so because one "wrong" is enough to eliminate the idea. But there are occasional opportunities in which we can actually demonstrate that a proposed concept is right, as in the case of observing individual atoms. This only became possible in very recent years, yet we accepted their existence for well over a century. Often, we live and work with a concept that is useful and consistent with all we can observe but that we cannot *prove* is correct. That, for many years, was the situation with the concept that matter consists of atoms. And that condition led to the realization that much of science evolves by depending on our ability to show that an idea or concept is wrong and inconsistent with what we can observe, while we must live with concepts that are indeed consistent with all we can observe but that we are unable to prove to be correct. And sometimes we eventually do develop means to test those ideas in a positive way, not just in the negative way of seeing whether they are wrong, or inconsistent with what we observe.

How Science Evolves

Sciences and scientific "laws" can never be thought of as permanent, inviolate statements about nature. We saw how our concept of the nature of heat emerged from the controversy over whether heat is a fluid, "caloric," or the result of the motion of the elementary particles in matter. We saw the difficulties and tensions associated with the emergence of the concept of energy, and we now can look back to see how that concept has changed and expanded as we have learned more and more about it. Until Einstein's theory of relativity, energy and mass were "known" to be two totally different manifestations of nature, but when the relation that showed that mass is indeed *a form of energy*, related to energy itself by the famous, simple expression $E = mc^2$, we had to expand our concept of the various forms that energy could take. "Atomic energy" is, literally, energy associated with the conversion of mass into other forms of energy. In a nuclear reactor, relatively unstable nuclei of high mass, typically of the uranium isotope U^{235}, break apart, and the sum of the masses of the fragments is slightly smaller than the mass of the original uranium. That loss of mass reappears typically as heat, which we use to generate electricity.

The concept of energy is undergoing another substantive expansion just as this is being written. We now know that there is something pervasive in the universe that exerts gravitational force, something whose nature we do not yet understand but whose presence is now well established, something we call "dark matter." Because it does exert observable, even very large gravitational effects, it must have some of the properties of mass and hence must be another form of energy. Its properties, revealed by its gravitational effects on galaxies, tell us much about where it is, but we still know very little about what it is. Beyond the puzzle of dark matter, and even more puzzling and less understood, is that there must be another form of energy—something, whatever it turns out to be, that is responsible for the accelerating expansion of the observable universe. That "something" goes by the name "dark energy," which merely acknowledges that it must be included with all the other manifestations of nature that we call "energy." So far as we know now, the one common property that we can associate with the many forms that energy can take is that of its *conservation.* And that is a truly amazing thing, perhaps one of the very few things about nature that we are inclined to accept, at least in this age, as a true, inviolable property of nature. A tantalizing question arises here, which the reader might ponder: is energy a discovery or an invention of the human mind?

We have seen Newton's mechanics, long thought to be totally fundamental, transformed into a useful description of what we experience in our everyday lives, but, nevertheless, a consequence of the more fundamental quantum mechanics. We are seeing thermodynamics transformed from a universal doctrine into a useful representation valid for large enough systems. We must be prepared, always, to learn that concepts that are useful at some scale or under some circumstances may be revealed to have more limited applicability than we originally thought—or even that those concepts, however plausible they might have been, were simply wrong, as with the concept of "caloric" as a fluid thought to constitute heat, and how that idea was discarded when people realized that heat is a consequence of random motion at the atomic level. A somewhat analogous idea was a widely accepted notion that fire was caused by a substance called "phlogiston," and that it was the release of phlogiston that was literally the flame. That concept, of course, was discarded when we came to recognize that burning in air is a consequence of a chemical reaction in which the combustible material, be it wood, paper, coal, or whatever else, combines with oxygen. But it was not until people found that in burning, materials *gained* weight, rather than losing it as the phlogiston theory predicted, that the true nature of combustion became known.

Let us take a broader, more distant view now, of science, of what it does, what it is, and how it evolves. Science consists of at least two kinds of knowledge, what we learn from observation and experiment and what we learn from creating interpretations of those experiments, the intellectual constructs we call theories. We can recognize, by looking at the history of thermodynamics, how our tools of observation have evolved, improved, become ever more powerful and sophisticated, from following the temperature associated with cannon-boring to the observation of Brownian motion in tiny particles, to the astronomical observations that reveal the presence of dark matter. And with the changing tools of observation, we have had to revise and reconstruct the theories we use to interpret these observations. Another vivid illustration is the history of the atomic theory of matter. As we saw, a kind of atomic theory arose in ancient Greece, through the ideas of Democritus and Lucretius, but the concept remained controversial, opposed to a concept that matter was infinitely divisible, until Einstein explained the random Brownian motion of colloidal particles as the result of random collisions of those particles with tinier particles that were still not visible then. But now, with the remarkable advances in the tools we can use, we can actually see and manipulate individual atoms. The atomic theory of matter is indisputable because we have unassailable evidence for the existence of these objects.

We can think of the evolution of each science in terms of the levels of understanding through which it passes with time, as our tools of observation develop and become more powerful, and our theories advance to enable us to understand those observations. Thus thermodynamics, and all other areas of science, are never static, locked combinations of observation and interpretation; they are constantly changing, becoming ever broader in the sense of encompassing new kinds of experience. Newton's physics remains valid in the domain for which it was intended, the mechanics of everyday experience. Quantum mechanics, in a sense, displaced it but actually left it valid and in a sense deepened in its proper domain, while enabling us to understand phenomena at the molecular level where the Newtonian picture fails to account for the observations.

An analogous pattern is apparent in thermodynamics, where we see the evolving, broadening nature arising not only from new kinds of observations but also from asking new questions. We can trace its history by following the questions that stimulated conceptual advances. Perhaps the first key question was the basis of Carnot's work, "How can we determine the best possible (meaning most efficient) performance of a heat-driven machine?" But there were other deep questions that followed: "What is energy and what are its inherent characteristics?" "What gives the directionality to evolution in time?" Then, more recently, "How can we extend the tools of traditional thermodynamics, based on equilibrium and idealized systems and processes, to more realistic situations, systems not in equilibrium, and systems operating at nonzero rates?" And "How can we extend thermodynamics to systems such as galaxies, whose energy depends on the products of masses of interacting bodies?" And still further, "How can we understand why some very powerful, general concepts of thermodynamics lose their validity when we try to apply those concepts to very small systems?" Thermodynamics is, in a sense, a science of things made of very many elements, not a means to treat and understand tiny things consisting of five or ten or a hundred atoms. We can expect thermodynamics, and all our other sciences, to continue to evolve, to explain newly observed phenomena and to answer new, ever more challenging questions. There probably will be no end to the evolving nature of science; it will continue to grow, to be ever more encompassing, to enable us to understand ever more about the universe, but we can see no end to this process.

Thus we see, partly through the evolution of thermodynamics, that our concepts of nature and what we think of as the substantive knowledge contained in a science is never firm and fixed. There have even been situations in which the concept of conservation of energy has been challenged. Until now, it has withstood those challenges, but we cannot say with total certainty that there are no limits to its validity. Science will continue to evolve as long as we can find new ways to explore the universe and whatever we find in it.

Index